American Merchant Ships and Sailors

Willis J. Abbot

Alpha Editions

This edition published in 2024

ISBN : 9789366388342

Design and Setting By
Alpha Editions
www.alphaedis.com
Email - info@alphaedis.com

As per information held with us this book is in Public Domain.
This book is a reproduction of an important historical work. Alpha Editions uses the best technology to reproduce historical work in the same manner it was first published to preserve its original nature. Any marks or number seen are left intentionally to preserve its true form.

Contents

PREFACE .... - 1 -
CHAPTER I. .... - 2 -
CHAPTER II. .... - 35 -
CHAPTER III .... - 57 -
CHAPTER IV .... - 76 -
CHAPTER V .... - 98 -
CHAPTER VI. .... - 122 -
CHAPTER VII. .... - 148 -
CHAPTER VIII .... - 166 -
CHAPTER IX .... - 192 -
CHAPTER X .... - 217 -

PREFACE

In an earlier series of books the present writer told the story of the high achievements of the men of the United States Navy, from the day of Paul Jones to that of Dewey, Schley, and Sampson. It is a record Americans may well regard with pride, for in wars of defense or offense, in wars just or unjust, the American blue jacket has discharged the duty allotted to him cheerfully, gallantly, and efficiently.

But there are triumphs to be won by sea and by land greater than those of war, dangers to be braved, more menacing than the odds of battle. It was a glorious deed to win the battle of Santiago, but Fulton and Ericsson influenced the progress of the world more than all the heroes of history. The daily life of those who go down to the sea in ships is one of constant battle, and the whaler caught in the ice-pack is in more direful case than the blockaded cruiser; while the captain of the ocean liner, guiding through a dense fog his colossal craft freighted with two thousand human lives, has on his mind a weightier load of responsibility than the admiral of the fleet.

In all times and ages, the deeds of the men who sail the deep as its policemen or its soldiery have been sung in praise. It is time for chronicle of the high courage, the reckless daring, and oftentimes the noble self-sacrifice of those who use the Seven Seas to extend the markets of the world, to bring nations nearer together, to advance science, and to cement the world into one great interdependent whole.

WILLIS JOHN ABBOT.
Ann Arbor, Mich., May 1, 1902.

CHAPTER I.

THE AMERICAN SHIP AND THE AMERICAN SAILOR—NEW ENGLAND'S LEAD ON THE OCEAN—THE EARLIEST AMERICAN SHIP-BUILDING—HOW THE SHIPYARDS MULTIPLIED—LAWLESS TIMES ON THE HIGH SEAS—SHIP-BUILDING IN THE FORESTS AND ON THE FARM—SOME EARLY TYPES—THE COURSE OF MARITIME TRADE—THE FIRST SCHOONER AND THE FIRST FULL-RIGGED SHIP—JEALOUSY AND ANTAGONISM OF ENGLAND—THE PEST OF PRIVATEERING—ENCOURAGEMENT FROM CONGRESS—THE GOLDEN DAYS OF OUR MERCHANT MARINE—FIGHTING CAPTAINS AND TRADING CAPTAINS—GROUND BETWEEN FRANCE AND ENGLAND—CHECKED BY THE WARS—SEALING AND WHALING—INTO THE PACIFIC—HOW YANKEE BOYS MOUNTED THE QUARTER-DECK—SOME STORIES OF EARLY SEAMEN—THE PACKETS AND THEIR EXPLOITS.

When the Twentieth Century opened, the American sailor was almost extinct. The nation which, in its early and struggling days, had given to the world a race of seamen as adventurous as the Norse Vikings had, in the days of its greatness and prosperity turned its eyes away from the sea and yielded to other people the mastery of the deep. One living in the past, reading the newspapers, diaries and record-books of the early days of the Nineteenth Century, can hardly understand how an occupation which played so great a part in American life as seafaring could ever be permitted to decline. The dearest ambition of the American boy of our early national era was to command a clipper ship—but how many years it has been since that ambition entered into the mind of young America! In those days the people of all the young commonwealths from Maryland northward found their interests vitally allied with maritime adventure. Without railroads, and with only the most wretched excuses for post-roads, the States were linked together by the sea; and coastwise traffic early began to employ a considerable number of craft and men. Three thousand miles of ocean separated Americans from the market in which they must sell their produce and buy their luxuries. Immediately upon the settlement of the seaboard the Colonists themselves took up this trade, building and manning their own vessels and speedily making their way into every nook and corner of Europe. We, who have seen, in the last quarter of the Nineteenth Century, the American flag the rarest of all ensigns to be met on the water, must regard with equal admiration and wonder the zeal for maritime adventure that made the infant nation of 1800 the second seafaring people in point of number of vessels, and second to none in energy and enterprise.

THE SHALLOP

New England early took the lead in building ships and manning them, and this was but natural since her coasts abounded in harbors; navigable streams ran through forests of trees fit for the ship-builder's adze; her soil was hard and obdurate to the cultivator's efforts; and her people had not, like those who settled the South, been drawn from the agricultural classes. Moreover, as I shall show in other chapters, the sea itself thrust upon the New Englanders its riches for them to gather. The cod-fishery was long pursued within a few miles of Cape Ann, and the New Englanders had become well habituated to it before the growing scarcity of the fish compelled them to seek the teeming waters of Newfoundland banks. The value of the whale was first taught them by great carcasses washed up on the shore of Cape Cod, and for years this gigantic game was pursued in open boats within sight of the coast. From neighborhood seafaring such as this the progress was easy to coasting voyages, and so to Europe and to Asia.

There is some conflict of historians over the time and place of the beginning of ship-building in America. The first vessel of which we have record was the "Virginia," built at the mouth of the Kennebec River in 1608, to carry home a discontented English colony at Stage Island. She was a two-master of 30 tons burden. The next American vessel recorded was the Dutch "yacht" "Onrest," built at New York in 1615. Nowadays sailors define a yacht as a vessel that carries no cargo but food and champagne, but the "Onrest" was not a yacht of this type. She was of 16 tons burden, and this small size explains her description.

The first ship built for commercial purposes in New England was "The Blessing of the Bay," a sturdy little sloop of 60 tons. Fate surely designed to give a special significance to this venture, for she was owned by John Winthrop, the first of New England statesmen, and her keel was laid on the Fourth of July, 1631—a day destined after the lapse of one hundred and forty-five years to mean much in the world's calendar. Sixty tons is not an awe-inspiring register. The pleasure yacht of some millionaire stock-jobber to-day will be ten times that size, while 20,000 tons has come to be an every-day register for an ocean vessel; but our pleasure-seeking "Corsairs," and our castellated "City of New York" will never fill so big a place in history as this little sloop, the size of a river lighter, launched at Mistick, and straightway dispatched to the trade with the Dutch at New Amsterdam. Long before her time, however, in 1526, the Spanish adventurer, Lucas Vasquez de Ayllon, losing on the coast of Florida a brigantine out of the squadron of three ships which formed his expedition, built a small craft called a gavarra to replace it.

From that early Fourth of July, for more than two hundred years shipyards multiplied and prospered along the American coast. The Yankees, with their racial adaptability, which long made them jacks of all trades and good at all, combined their shipbuilding with other industries, and to the hurt of neither. Early in 1632, at Richmond Island, off the coast of Maine, was built what was probably the first regular packet between England and America. She carried to the old country lumber, fish, furs, oil, and other colonial products, and brought back guns, ammunition, and liquor—not a fortunate exchange. Of course meanwhile English, Dutch, and Spanish ships were trading to the colonies, and every local essay in shipbuilding meant competition with old and established ship-yards and ship owners. Yet the industry throve, not only in the considerable yards established at Boston and other large towns, but in a small way all along the coast. Special privileges were extended to ship-builders. They were exempt from military and other public duties. In 1636 the "Desire," a vessel of 120 tons, was built at Marblehead, the largest to that time. By 1640 the port records of European ports begin to show the clearings of American-built vessels.

THE KETCH

In those days of wooden hulls and tapering masts the forests of New England were the envy of every European monarch ambitious to develop a navy. It was a time, too, of greater naval activity than the world had ever seen—though but trivial in comparison with the present expenditures of Christian nations for guns and floating steel fortresses. England, Spain, Holland, and France were struggling for the control of the deep, and cared little for considerations of humanity, honor, or honesty in the contest. The tall, straight pines of Maine and New Hampshire were a precious possession for England in the work of building that fleet whose sails were yet to whiten the ocean, and whose guns, under Drake and Rodney, were to destroy successfully the maritime prestige of the Dutch and the Spaniards. Sometimes a colony, seeking royal favor, would send to the king a present of these pine timbers, 33 to 35 inches in diameter, and worth £95 to £115 each. Later the royal mark, the "broad arrow," was put on all white pines 24 inches in diameter 3 feet from the ground, that they might be saved for masts. It is, by the way, only about fifteen years since our own United States Government has disposed of its groves of live oaks, that for nearly a century were preserved to furnish oaken knees for navy vessels.

"THE BROAD ARROW WAS PUT ON ALL WHITE PINES 24 INCHES IN DIAMETER"

The great number of navigable streams soon led to shipbuilding in the interior. It was obviously cheaper to build the vessel at the edge of the forest, where all the material grew ready to hand, and sail the completed craft to the seaboard, than to first transport the material thither in the rough. But American resourcefulness before long went even further. As the forests receded from the banks of the streams before the woodman's axe, the shipwrights followed. In the depths of the woods, miles perhaps from water, snows, pinnaces, ketches, and sloops were built. When the heavy snows of winter had fallen, and the roads were hard and smooth, runners were laid under the little ships, great teams of oxen—sometimes more than one hundred yoke—were attached, and the craft dragged down to the river, to lie there on the ice until the spring thaw came to gently let it down into its proper element. Many a farmer, too, whose lands sloped down to a small harbor, or stream, set up by the water side the frame of a vessel, and worked patiently at it during the winter days when the flinty soil repelled the plough and farm work was stopped. Stout little craft were thus put together, and sometimes when the vessel was completed the farmer-builder took his place at the helm and steered her to the fishing banks, or took her through Hell Gate to the great and thriving city of New York. The world has never seen a more amphibious populace.

"THE FARMER-BUILDER TOOK HIS PLACE AT THE HELM"

The cost of the little vessels of colonial times we learn from old letters and accounts to have averaged four pounds sterling to the ton. Boston, Charleston, Salem, Ipswich, Salisbury, and Portsmouth were the chief building places in Massachusetts; New London in Connecticut, and Providence in Rhode Island. Vessels of a type not seen to-day made up the greater part of the New England fleet. The ketch, often referred to in early annals, was a two-master, sometimes rigged with lanteen sails, but more often with the foremast square-rigged, like a ship's foremast, and the mainmast like the mizzen of a modern bark, with a square topsail surmounting a fore-and-aft mainsail. The foremast was set very much aft—often nearly amidships. The snow was practically a brig, carrying a fore-and-aft sail on the mainmast, with a square sail directly above it. A pink was rigged like a schooner, but without a bowsprit or jib. For the fisheries a multitude of smaller types were constructed—such as the lugger, the shallop, the sharpie, the bug-eye, the smack. Some of these survive to the present day, and in many cases the name has passed into disuse, while the type itself is now and then to be met with on our coasts.

The importance of ship-building as a factor in the development of New England did not rest merely upon the use of ships by the Americans alone. That was a day when international trade was just beginning to be understood and pushed, and every people wanted ships to carry their goods to foreign lands and bring back coveted articles in exchange. The New England vessel seldom made more than two voyages across the Atlantic without being snapped up by some purchaser beyond seas. The ordinary course was for the new craft to load with masts or spars, always in demand, or with fish; set sail for a promising market, dispose of her cargo, and take freight for England. There she would be sold, her crew making their way home in other ships, and her purchase money expended in articles needed in the colonies. This was the ordinary practice, and with vessels sold abroad so soon after their completion the shipyards must have been active to have fitted out, as the records show, a fleet of fully 280 vessels for Massachusetts alone by 1718. Before this time, too, the American shipwrights had made such progress in the mastery of their craft that they were building ships for the royal navy. The "Falkland," built at Portsmouth about 1690, and carrying 54 guns, was the earliest of these, but after her time corvettes, sloops-of-war, and frigates were launched in New England yards to fight for the king. It was good preparation for building those that at a later date should fight against him.

Looking back over the long record of American maritime progress, one cannot but be impressed with the many and important contributions made by Americans—native or adopted—to marine architecture. To an American citizen, John Ericsson, the world owes the screw propeller. Americans sent the first steamship across the ocean—the "Savannah," in 1819. Americans, engaged in a fratricidal war, invented the ironclad in the "Monitor" and the "Merrimac," and, demonstrating the value of iron ships for warfare, sounded the knell of wooden ships for peaceful trade. An American first demonstrated the commercial possibilities of the steamboat, and if history denies to Fulton entire precedence with his "Clermont," in 1807, it may still be claimed for John Fitch, another American, with his imperfect boat on the Delaware in 1787. But perhaps none of these inventions had more homely utility than the New England schooner, which had its birth and its christening at Gloucester in 1713. The story of its naming is one of the oldest in our marine folk-lore.

"See how she schoons!" cried a bystander, coining a verb to describe the swooping slide of the graceful hull down the ways into the placid water.

SCHOONER-RIGGED SHARPIE

- 8 -

"A schooner let her be!" responded the builder, proud of his handiwork, and ready to seize the opportunity to confer a novel title upon his novel creation. Though a combination of old elements, the schooner was in effect a new design. Barks, ketches, snows, and brigantines carried fore-and-aft rigs in connection with square sails on either mast, but now for the first time two masts were rigged fore and aft, and the square sails wholly discarded. The advantages of the new rig were quickly discovered. Vessels carrying it were found to sail closer to the wind, were easier to handle in narrow quarters, and—what in the end proved of prime importance—could be safely manned by smaller crews. With these advantages the schooner made its way to the front in the shipping lists. The New England shipyards began building them, almost to the exclusion of other types. Before their advance brigs, barks, and even the magnificent full-rigged ship itself gave way, until now a square-rigged ship is an unusual spectacle on the ocean. The vitality of the schooner is such that it bids fair to survive both of the crushing blows dealt to old-fashioned marine architecture—the substitution of metal for wood, and of steam for sails. To both the schooner adapted itself. Extending its long, slender hull to carry four, five, and even seven masts, its builders abandoned the stout oak and pine for molded iron and later steel plates, and when it appeared that the huge booms, extending the mighty sails, were difficult for an ordinary crew to handle, one mast, made like the rest of steel, was transformed into a smokestack—still bearing sails—a donkey engine was installed in the hold, and the booms went aloft, or the anchor rose to the peak to the tune of smoky puffing instead of the rhythmical chanty songs of the sailors. So the modern schooner, a very leviathan of sailing craft, plows the seas, electric-lighted, steering by steam, a telephone system connecting all parts of her hull—everything modern about her except her name. Not as dignified, graceful, and picturesque as the ship perhaps—but she lasts, while the ship disappears.

But to return to the colonial shipping. Boston soon became one of the chief building centers, though indeed wherever men were gathered in a seashore village ships were built. Winthrop, one of the pioneers in the industry, writes: "The work was hard to accomplish for want of money, etc., but our shipwrights were content to take such pay as the country could make," and indeed in the old account books of the day we can read of very unusual payments made for labor, as shown, for example, in a contract for building a ship at Newburyport in 1141, by which the owners were bound to pay "£300 in cash, £300 by orders on good shops in Boston; two-thirds money; four hundred pounds by orders up the river for tim'r and plank, ten bbls. flour, 50 pounds weight of loaf sugar, one bagg of cotton wool, one hund. bushels of corn in the spring; one hhd. of Rum, one hundred weight of cheese * * * whole am't of price for vessel £3000 lawful money."

By 1642 they were building good-sized vessels at Boston, and the year following was launched the first full-rigged ship, the "Trial," which went to Malaga, and brought back "wine, fruit, oil, linen and wool, which was a great advantage to the country, and gave encouragement to trade." A year earlier there set out the modest forerunner of our present wholesale spring pilgrimages to Europe. A ship set sail for London from Boston "with many passengers, men of chief rank in the country, and great store of beaver. Their adventure was very great, considering the doubtful estate of affairs of England, but many prayers of the churches went with them and followed after them."

By 1698 Governor Bellomont was able to say of Boston alone, "I believe there are more good vessels belonging to the town of Boston than to all Scotland and Ireland." Thereafter the business rapidly developed, until in a map of about 1730 there are noted sixteen shipyards. Rope walks, too, sprung up to furnish rigging, and presently for these Boston was a centre. Another industry, less commendable, grew up in this as in other shipping centres. Molasses was one of the chief staples brought from the West Indies, and it came in quantities far in excess of any possible demand from the colonial sweet tooth. But it could be made into rum, and in those days rum was held an innocent beverage, dispensed like water at all formal gatherings, and used as a matter of course in the harvest fields, the shop, and on the deck at sea. Moreover, it had been found to have a special value as currency on the west coast of Africa. The negro savages manifested a more than civilized taste for it, and were ready to sell their enemies or their friends, their sons, fathers, wives, or daughters into slavery in exchange for the fiery fluid. So all New England set to turning the good molasses into fiery rum, and while the slave trade throve abroad the rum trade prospered at home.

Of course the rapid advance of the colonies in shipbuilding and in maritime trade was not regarded in England with unqualified pride. The theory of that day—and one not yet wholly abandoned—was that a colony was a mine, to be worked for the sole benefit of the mother country. It was to buy its goods in no other market. It was to use the ships of the home government alone for its trade across seas. It must not presume to manufacture for itself articles which merchants at home desired to sell. England early strove to impress such trade regulations upon the American colonies, and succeeded in embarrassing and handicapping them seriously, although evasions of the navigation laws were notorious, and were winked at by the officers of the crown. The restrictions were sufficiently burdensome, however, to make the ship-owners and sailors of 1770 among those most ready and eager for the revolt against the king.

The close of the Revolution found American shipping in a reasonably prosperous condition. It is true that the peaceful vocation of the seamen had been interrupted, all access to British ports denied them, and their voyages to Continental markets had for six years been attended by the ever-present risk of capture and condemnation. But on the other hand, the war had opened the way for privateering, and out of the ports of Massachusetts, Rhode Island and Connecticut the privateers swarmed like swallows from a chimney at dawn. To the adventurous and not over-scrupulous men who followed it, privateering was a congenial pursuit—so much so, unhappily, that when the war ended, and a treaty robbed their calling of its guise of lawfulness, too many of them still continued it, braving the penalties of piracy for the sake of its gains. But during the period of the Revolution privateering did the struggling young nation two services—it sorely harassed the enemy, and it kept alive the seafaring zeal and skill of the New Englanders.

For a time it seemed that not all this zeal and skill could replace the maritime interests where they were when the Revolution began. For most people in the colonies independence meant a broader scope of activity—to the shipowner and sailor it meant new and serious limitations. England was still engaged in the effort to monopolize ocean traffic by the operation of tariffs and navigation laws. New England having become a foreign nation, her ships were denied admittance to the ports of the British West Indies, with which for years a nourishing trade had been conducted. Lumber, corn, fish, live stock, and farm produce had been sent to the islands, and coffee, sugar, cotton, rum, and indigo brought back. This commerce, which had come to equal £3,500,000 a year, was shut off by the British after American independence, despite the protest of Pitt, who saw clearly that the West Indians would suffer even more than the Americans. Time showed his wisdom. Terrible sufferings came upon the West Indies for lack of the supplies they had been accustomed to import, and between 1780 and 1787 as many as 15,000 slaves perished from starvation.

Another cause held the American merchant marine in check for several years succeeding the declaration of peace. If there be one interest which must have behind it a well-organized, coherent national government, able to protect it and to enforce its rights in foreign lands, it is the shipping interest. But American ships, after the Treaty of Paris, hailed from thirteen independent but puny States. They had behind them the shadow of a confederacy, but no substance. The flags they carried were not only not respected in foreign countries—they were not known. Moreover, the States were jealous of each other, possessing no true community of interest, and each seeking advantage at the expense of its neighbors. They were already beginning to adopt among themselves the very tactics of harassing and crippling navigation laws which caused the protest against Great Britain. This "Critical Period of American History," as Professor Fiske calls it, was indeed a critical period for American shipping.

The new government, formed under the Constitution, was prompt to recognize the demands of the shipping interests upon the country. In the very first measure adopted by Congress steps were taken to encourage American shipping by differential duties levied on goods imported in American and foreign vessels. Moreover, in the tonnage duties imposed by Congress an advantage of almost 50 per cent. was given ships built in the United States and owned abroad. Under this stimulus the shipping interests throve, despite hostile legislation in England, and the disordered state of the high seas, where French and British privateers were only a little less predatory than Algerian corsairs or avowed pirates. It was at this early day that Yankee skippers began making those long voyages that are hardly paralleled to-day when steamships hold to a single route like a trolley car between two towns. The East Indies was a favorite trading point. Carrying a cargo suited to the needs of perhaps a dozen different peoples, the vessel would put out from Boston or Newport, put in at Madeira perhaps, or at some West Indian port, dispose of part of its cargo, and proceed, stopping again and again on its way, and exchanging its goods for money or for articles thought to be more salable in the East Indies. Arrived there, all would be sold, and a cargo of tea, coffee, silks, spices, nankeen cloth, sugar, and other products of the country taken on. If these goods did not prove salable at home the ship would make yet another voyage and dispose of them at Hamburg or some other Continental port. In 1785 a Baltimore ship showed the Stars and Stripes in the Canton River, China. In 1788 the ship "Atlantic," of Salem, visited Bombay and Calcutta. The effect of being barred from British ports was not, as the British had expected, to put an abrupt end to American maritime enterprise. It only sent our hardy seamen on longer voyages, only brought our merchants into touch with the commerce of the most distant lands. Industry, like men, sometimes thrives upon obstacles.

"AFTER A BRITISH LIEUTENANT HAD PICKED THE BEST OF HER CREW"

For twenty-five years succeeding the adoption of the Constitution the maritime interest—both shipbuilding and shipowning—thrived more, perhaps, than any other gainful industry pursued by the Americans. Yet it was a time when every imaginable device was employed to keep our people out of the ocean-carrying trade. The British regulations, which denied us access to their ports, were imitated by the French. The Napoleonic wars came on, and the belligerents bombarded each other with orders in council and decrees that fell short of their mark, but did havoc among neutral merchantmen. To the ordinary perils of the deep the danger of capture—lawful or unlawful—by cruiser or privateer, was always to be added. The British were still enforcing their so-called "right of search," and many an American ship was left short-handed far out at sea, after a British naval lieutenant had picked the best of her crew on the pretense that they were British subjects. The superficial differences between an American and an Englishman not being as great as those between an albino and a Congo black, it is not surprising that the boarding officer should occasionally make mistakes—particularly when his ship was in need of smart, active sailors. Indeed, in those years the civilized—by which at that period was meant the warlike—nations were all seeking sailors. Dutch, Spanish, French, and English were eager for men to man their fighting ships; hired them when they could, and stole them when they must. It was the time of the press gang, and the day when sailors carried as a regular part of their kit an outfit of women's clothing in which to escape if the word were passed that "the press is hot to-night." The United States had never to resort to impressment to fill its navy ships' companies, a fact perhaps due chiefly to the small size of its navy in comparison with the seafaring population it had to draw from.

As for the American merchant marine, it was full of British seamen. Beyond doubt inducements were offered them at every American port to desert and ship under the Stars and Stripes. In the winter of 1801 every British ship visiting New York lost the greater part of its crew. At Norfolk the entire crew of a British merchantman deserted to an American sloop-of-war. A lively trade was done in forged papers of American citizenship, and the British naval officer who gave a boat-load of bluejackets shore leave at New York was liable to find them all Americans when their leave was up. Other nations looked covetously upon our great body of able-bodied seamen, born within sound of the swash of the surf, nurtured in the fisheries, able to build, to rig, or to navigate a ship. They were fighting sailors, too, though serving only in the merchant marine. In those days the men that went down to the sea in ships had to be prepared to fight other antagonists than Neptune and Æolus. All the ships went armed. It is curious to read in old annals of the number of cannon carried by small merchantmen. We find the "Prudent Sarah" mounting 10 guns; the "Olive Branch," belied her peaceful name with 3, while the pink "Friendship" carried 8. These years, too, were the privateers' harvest time. During the Revolution the ships owned by one Newburyport merchant took 23,360 tons of shipping and 225 men, the prizes with their cargoes selling for $3,950,000. But of the size and the profits of the privateering business more will be said in the chapter devoted to that subject. It is enough to note here that it made the American merchantman essentially a fighting man.

The growth of American shipping during the years 1794-1810 is almost incredible in face of the obstacles put in its path by hostile enactments and the perils of the war. In 1794 United States ships, aggregating 438,863 tons, breasted the waves, carrying fish and staves to the West Indies, bringing back spices, rum, cocoa, and coffee. Sometimes they went from the West Indies to the Canaries, and thence to the west coast of Africa, where very valuable and very pitiful cargoes of human beings, whose black skins were thought to justify their treatment as dumb beasts of burden, were shipped. Again the East Indies opened markets for buying and selling both. But England and almost the whole of Western Europe were closed.

It is not possible to understand the situation in which the American sailor and shipowner of that day was placed, without some knowledge of the navigation laws and belligerent orders by which the trade was vexed. In 1793 the Napoleonic wars began, to continue with slight interruptions until 1815. France and England were the chief contestants, and between them American shipping was sorely harried. The French at first seemed to extend to the enterprising Americans a boon of incalculable value to the maritime interest, for the National Convention promulgated a decree giving to neutral ships— practically to American ships, for they were the bulk of the neutral shipping—the rights of French ships. Overjoyed by this sudden opening of a rich market long closed, the Yankee barks and brigs slipped out of the New England harbors in schools, while the shipyards rung with the blows of the hammers, and the forest resounded with the shouts of the woodsmen getting out ship-timbers. The ocean pathway to the French West Indies was flecked with sails, and the harbors of St. Kitts, Guadaloupe, and Martinique were crowded. But this bustling trade was short-lived. The argosies that set forth on their peaceful errand were shattered by enemies more dreaded than wind or sea. Many a ship reached the port eagerly sought only to rot there; many a merchant was beggared, nor knew what had befallen his hopeful venture until some belated consular report told of its condemnation in some French or English admiralty court.

EARLY TYPE OF SMACK

For England met France's hospitality with a new stroke at American interests. The trade was not neutral, she said. France had been forced to her concession by war. Her people were starving because the vigilance of British cruisers had driven French cruisers from the seas, and no food could be imported. To permit Americans to purvey food for the French colonies would clearly be to undo the good work of the British navy. Obviously food was contraband of war. So all English men-of-war were ordered to seize French goods on whatever ship found; to confiscate cargoes of wheat, corn, or fish bound for French ports as contraband, and particularly to board all American merchantmen and scrutinize the crews for English-born sailors. The latter injunction was obeyed with peculiar zeal, so that the State Department had evidence that at one time, in 1806, there were as many as 6000 American seamen serving unwillingly in the British navy.

France, meanwhile, sought retaliation upon England at the expense of the Americans. The United States, said the French government, is a sovereign nation. If it does not protect its vessels against unwarrantable British aggressions it is because the Americans are secretly in league with the British. France recognizes no difference between its foes. So it is ordered that any American vessel which submitted to visitation and search from an English vessel, or paid dues in a British port, ceased to be neutral, and became subject to capture by the French. The effect of these orders and decrees was simply that any American ship which fell in with an English or French man-of-war or privateer, or was forced by stress of weather to seek shelter in an English or French port, was lost to her owners. The times were rude, evidence was easy to manufacture, captains were rapacious, admiralty judges were complaisant, and American commerce was rich prey. The French West Indies fell an easy spoil to the British, and at Martinique and Basseterre American merchantmen were caught in the harbor. Their crews were impressed, their cargoes, not yet discharged, seized, the vessels themselves wantonly destroyed or libelled as prizes. Nor were passengers exempt from the rigors of search and plunder. The records of the State Department and the rude newspapers of the time are full of the complaints of shipowners, passengers, and shipping merchants. The robbery was prodigious in its amount, the indignity put upon the nation unspeakable. And yet the least complaint came from those who suffered most. The New England seaport towns were filled with idle seamen, their harbors with pinks, schooners, and brigs, lying lazily at anchor. The sailors, with the philosophy of men long accustomed to submit themselves to nature's moods and the vagaries of breezes, cursed British and French impartially, and joined in the general depression and idleness of the towns and counties dependent on their activity.

It was about this period (1794) that the American navy was begun; though, curiously enough, its foundation was not the outcome of either British or French depredations, but of the piracies of the Algerians. That fierce and predatory people had for long years held the Mediterranean as a sort of a private lake into which no nation might send its ships without paying tribute. With singular cowardice, all the European peoples had acquiesced in this conception save England alone. The English were feared by the Algerians, and an English pass—which tradition says the illiterate Corsairs identified by measuring its enscrolled border, instead of by reading—protected any vessel carrying it. American ships, however, were peculiarly the prey of the Algerians, and many an American sailor was sold by them into slavery until Decatur and Rodgers in 1805 thrashed the piratical states of North Africa into recognition of American power. In 1794, however, the Americans were not eager for war, and diplomats strove to arrange a treaty which would protect American shipping, while Congress prudently ordered the beginning of six frigates, work to be stopped if peace should be made with the Dey. The treaty—not one very honorable to us—was indeed made some months later, and the frigates long remained unfinished.

It has been the fashion of late years to sneer at our second war with England as unnecessary and inconclusive. But no one who studies the records of the life, industry, and material interests of our people during the years between the adoption of the Constitution and the outbreak of that war can fail to wonder that it did not come sooner, and that it was not a war with France as well as England. For our people were then essentially a maritime people. Their greatest single manufacturing industry was ship-building. The fisheries—whale, herring, and cod—employed thousands of their men and supported more than one considerable town. The markets for their products lay beyond seas, and for their commerce an undisputed right to the peaceful passage of the ocean was necessary. Yet England and France, prosecuting their own quarrel, fairly ground American shipping as between two millstones. Our sailors were pressed, our ships seized, their cargoes stolen, under hollow forms of law. The high seas were treated as though they were the hunting preserves of these nations and American ships were quail and rabbits. The London "Naval Chronicle" at that time, and for long after, bore at the head of its columns the boastful lines:

> "The sea and waves are Britain's broad domain,
>
> And not a sail but by permission spreads."

And France, while vigorously denying the maxim in so far as it related to British domination, was not able to see that the ocean could be no one nation's domain, but must belong equally to all. It was the time when the French were eloquently discoursing of the rights of man; but they did not appear to regard the peaceful navigation of the ocean as one of those rights; they were preaching of the virtues of the American republic, but their rulers issued orders and decrees that nearly brought the two governments to the point of actual war. But the very fact that France and England were almost equally arrogant and aggressive delayed the formal declaration of hostilities. Within the United States two political parties—the Federalists and the Republicans—were struggling for mastery. The one defended, though half-heartedly, the British, and demanded drastic action against the French spoliators. The other denounced British insolence and extolled our ancient allies and brothers in republicanism, the French. While the politicians quarreled the British stole our sailors and the French stole our ships. In 1798 our, then infant, navy gave bold resistance to the French ships, and for a time a quasi-war was waged on the ocean, in which the frigates "Constitution" and "Constellation" laid the foundation for that fame which they were to finally achieve in the war with Great Britain in 1812. No actual war with France grew out of her aggressions. The Republicans came into power in the United States, and by diplomacy averted an actual conflict. But the American shipping interests suffered sadly meanwhile. The money finally paid by France as indemnity for her unwarranted spoliations lay long undivided in the United States Treasury, and the easy-going labor of urging and adjudicating French spoliation claims furnished employment to some generations of politicians after the despoiled seamen and shipowners had gone down into their graves.

In 1800 the whole number of American ships in foreign and coasting trades and the fisheries had reached a tonnage of 972,492. The growth was constant, despite the handicap resulting from the European wars. Indeed, it is probable that those wars stimulated American shipping more than the restrictive decrees growing out of them retarded it, for they at least kept England and France (with her allies) out of the active encouragement of maritime enterprise. But the vessels of that day were mere pigmies, and the extent of the trade carried on in them would at this time seem trifling. The gross exports and imports of the United States in 1800 were about $75,000,000 each. The vessels that carried them were of about 250 tons each, the largest attaining 400 tons. An irregular traffic was carried on along the coast, and it was 1801 before the first sloop was built to ply regularly on the Hudson between New York and Albany. She was of 100 tons, and carried passengers only. Sometimes the trip occupied a week, and the owner of the sloop established an innovation by supplying beds, provisions, and wines for his passengers. Between Boston and New York communication was still irregular, passengers waiting for cargoes. But small as this maritime interest now seems, more money was invested in it, and it occupied more men, than any other American industry, save only agriculture.

To this period belong such shipowners as William Gray, of Boston, who in 1809, though he had sixty great square-rigged ships in commission, nevertheless heartily approved of the embargo with which President Jefferson vainly strove to combat the outrages of France and England. Though the commerce of those days was world-wide, its methods—particularly on the bookkeeping side—were primitive. "A good captain," said Merchant Gray, "will sail with a load of fish to the West Indies, hang up a stocking in the cabin on arriving, put therein hard dollars as he sells fish, and pay out when he buys rum, molasses, and sugar, and hand in the stocking on his return in full of all accounts." The West Indies, though a neighboring market, were far from monopolizing the attention of the New England shipping merchants. Ginseng and cash were sent to China for silks and tea, the voyage each way, around the tempestuous Horn, occupying six months. In 1785 the publication of the journals of the renowned explorer, Captain Cook, directed the ever-alert minds of the New Englanders to the great herds of seal and sea-otters on the northwestern coast of the United States, and vessels were soon faring thither in pursuit of fur-bearing animals, then plentiful, but now bidding fair to become as rare as the sperm-whale. A typical expedition of this sort was that of the ship "Columbia," Captain Kendrick, and the sloop "Washington," Captain Gray, which sailed September 30, 1787, bound to the northwest coast and China. The merchant who saw his ships drop down the bay bound on such a voyage said farewell to them for a long time—perhaps forever. Years must pass before he could know whether the money he had invested, the cargo he had adventured, the stout ships he had dispatched, were to add to his fortune or to be at last a total loss. Perhaps for months he might be going about the wharves and coffee-houses, esteeming himself a man of substance and so held by all his neighbors, while in fact his all lay whitening in the surf on some far-distant Pacific atoll. So it was almost three years before news came back to Boston of these two ships; but then it was glorious, for then the "Federalist," of New York, came into port, bringing tidings that at Canton she had met the "Columbia," and had been told of the discovery by that vessel of the great river in Oregon to which her name had been given. Thus Oregon and Washington were given to the infant Union, the latter perhaps taking its name from the little sloop of 90 tons which accompanied the "Columbia" on her voyage. Six months later the two vessels reached Boston, and were greeted with salutes of cannon from the forts. They were the first American vessels to circumnavigate the globe. It is pleasant to note that a voyage which was so full of advantage to the nation was profitable to the owners. Thereafter an active trade was done with miscellaneous goods to the northwest Indians, skins and furs thence to the Chinese, and teas home. A typical outbound cargo in this trade was that of the "Atakualpa" in 1800. The vessel was of 218 tons, mounted eight guns, and was freighted with broadcloth, flannel, blankets, powder, muskets, watches, tools, beads, and looking-glasses. How great were the proportions that this trade speedily assumed may be judged from the fact that between June, 1800, and January, 1803, there were imported into China, in American vessels, 34,357 sea-otter skins worth on an average $18 to $20 each. Over a million sealskins were imported. In this trade were employed 80 ships and 9 brigs and schooners, more than half of them from Boston.

THE SNOW, AN OBSOLETE TYPE

- 19 -

Indeed, by the last decade of the eighteenth century Boston had become the chief shipping port of the United States. In 1790 the arrivals from abroad at that port were 60 ships, 7 snows, 159 brigs, 170 schooners, 59 sloops, besides coasters estimated to number 1,220 sail. In the *Independent Chronicle*, of October 27, 1791, appears the item: "Upwards of seventy sail of vessels sailed from this port on Monday last, for all parts of the world." A descriptive sketch, written in 1794 and printed in the Massachusetts Historical Society collections, says of the appearance of the water front at that time:

"There are eighty wharves and quays, chiefly on the east side of the town. Of these the most distinguished is Boston pier, or the Long Wharf, which extends from the bottom of State Street 1,743 feet into the harbor. Here the principal navigation of the town is carried on; vessels of all burdens load and unload; and the London ships generally discharge their cargoes.... The harbor of Boston is at this date crowded with vessels. It is reckoned that not less than 450 sail of ships, brigs, schooners, sloops, and small craft are now in this port."

New York and Baltimore, in a large way; Salem, Hull, Portsmouth, New London, New Bedford, New Haven, and a host of smaller seaports, in a lesser degree, joined in this prosperous industry. It was the great interest of the United States, and so continued, though with interruptions, for more than half a century, influencing the thought, the legislation, and the literature of our people. When Daniel Webster, himself a son of a seafaring State, sought to awaken his countrymen to the peril into which the nation was drifting through sectional dissensions and avowed antagonism to the national authority, he chose as the opening metaphor of his reply to Hayne the description of a ship, drifting rudderless and helpless on the trackless ocean, exposed to perils both known and unknown. The orator knew his audience. To all New England the picture had the vivacity of life. The metaphors of the sea were on every tongue. The story is a familiar one of the Boston clergyman who, in one of his discourses, described a poor, sinful soul drifting toward shipwreck so vividly that a sailor in the audience, carried away by the preacher's imaginative skill, cried out: "Let go your best bower anchor, or you're lost." In another church, which had its pulpit set at the side instead of at the end, as customary, a sailor remarked critically: "I don't like this craft; it has its rudder amidships."

At this time, and, indeed, for perhaps fifty years thereafter, the sea was a favorite career, not only for American boys with their way to make in the world, but for the sons of wealthy men as well. That classic of New England seamanship, "Two Years Before the Mast," was not written until the middle of the nineteenth century, and its author went to sea, not in search of wealth, but of health. But before the time of Richard Henry Dana, many a young man of good family and education—a Harvard graduate like him, perhaps—

bade farewell to a home of comfort and refinement and made his berth in a smoky, fetid forecastle to learn the sailor's calling. The sons of the great shipping merchants almost invariably made a few voyages—oftenest as supercargoes, perhaps, but not infrequently as common seamen. In time special quarters, midway between the cabin and the forecastle, were provided for these apprentices, who were known as the "ship's cousins." They did the work of the seamen before the mast, but were regarded as brevet officers. There was at that time less to engage the activities and arouse the ambitions of youth than now, and the sea offered the most promising career. Moreover, the trading methods involved, and the relations of the captain or other officers to the owners, were such as to spur ambition and promise profit. The merchant was then greatly dependent on his captain, who must judge markets, buy and sell, and shape his course without direction from home. So the custom arose of giving the captain—and sometimes other officers—an opportunity to carry goods of their own in the ship, or to share the owner's adventure. In the whaling and fishery business we shall see that an almost pure communism prevailed. These conditions attracted to the maritime calling men of an enterprising and ambitious nature—men to whom the conditions to-day of mere wage servitude, fixed routes, and constant dependence upon the cabled or telegraphed orders of the owner would be intolerable. Profits were heavy, and the men who earned them were afforded opportunities to share them. Ships were multiplying fast, and no really lively and alert seaman need stay long in the forecastle. Often they became full-fledged captains and part owners at the age of twenty-one, or even earlier, for boys went to sea at ages when the youngsters of equally prosperous families in these days would scarcely have passed from the care of a nurse to that of a tutor. Thomas T. Forbes, for example, shipped before the mast at the age of thirteen; was commander of the "Levant" at twenty; and was lost in the Canton River before he was thirty. He was of a family great in the history of New England shipping for a hundred years. Nathaniel Silsbee, afterwards United States Senator from Massachusetts, was master of a ship in the East India trade before he was twenty-one; while John P. Cushing at the age of sixteen was the sole—and highly successful—representative in China of a large Boston house. William Sturges, afterwards the head of a great world-wide trading house, shipped at seventeen, was a captain and manager in the China trade at nineteen, and at twenty-nine left the quarter-deck with a competence to establish his firm, which at one time controlled half the trade between the United States and China. A score of such successes might be recounted.

But the fee which these Yankee boys paid for introduction into their calling was a heavy one. Dana's description of life in the forecastle, written in 1840, holds good for the conditions prevailing for forty years before and forty after he penned it. The greeting which his captain gave to the crew of the brig "Pilgrim" was repeated, with little variation, on a thousand quarter-decks:

"Now, my men, we have begun a long voyage. If we get along well together we shall have a comfortable time; if we don't, we shall hay hell afloat. All you have to do is to obey your orders and do your duty like men—then you will fare well enough; if you don't, you will fare hard enough, I can tell you. If we pull together you will find me a clever fellow; if we don't, you will find me a bloody rascal. That's all I've got to say. Go below the larboard watch."

But the note of roughness and blackguardism was not always sounded on American ships. We find, in looking over old memoirs, that more than one vessel was known as a "religious ship"—though, indeed, the very fact that few were thus noted speaks volumes for the paganism of the mass. But the shipowners of Puritan New England not infrequently laid stress on the moral character of the men shipped. Nathaniel Ames, a Harvard graduate who shipped before the mast, records that on his first vessel men seeking berths even in the forecastle were ordered to bring certificates of good character from the clergyman whose church they had last attended. Beyond doubt, however, this was a most unusual requirement. More often the majority of the crew were rough, illiterate fellows, often enticed into shipping while under the influence of liquor, and almost always coming aboard at the last moment, much the worse for long debauches. The men of a better sort who occasionally found themselves unluckily shipped with such a crew, have left on record many curious stories of the way in which sailors, utterly unable to walk on shore or on deck for intoxication, would, at the word of command, spring into the rigging, clamber up the shrouds, shake out reefs, and perform the most difficult duties aloft.

THE BUG-EYE

Most of the things which go to make the sailor's lot at least tolerable nowadays, were at that time unknown. A smoky lamp swung on gimbals half-lighted the forecastle—an apartment which, in a craft of scant 400 tons, did not afford commodious quarters for a crew of perhaps a score, with their sea chests and bags. The condition of the fetid hole at the beginning of the voyage, with four or five apprentices or green hands deathly sick, the hardened seamen puffing out clouds of tobacco smoke, and perhaps all redolent of rum, was enough to disenchant the most ardent lover of the sea. The food, bad enough in all ages of seafaring, was, in the early days of our merchant marine, too often barely fit to keep life in men's bodies. The unceasing round of salt pork, stale beef, "duff," "lobscouse," doubtful coffee sweetened with molasses, and water, stale, lukewarm, and tasting vilely of the hogshead in which it had been stored, required sturdy appetites to make it even tolerable. Even in later days Frank T. Bullen was able to write: "I have often seen the men break up a couple of biscuits into a pot of coffee for their breakfast, and after letting it stand a minute or two, skim off the accumulated scum of vermin from the top—maggots, weevils, etc—to the extent of a couple of tablespoonsful, before they could shovel the mess into their craving stomachs."

It may be justly doubted whether history has ever known a race of men so hardy, so self-reliant, so adaptable to the most complex situations, so determined to compel success, and so resigned in the presence of inevitable failure, as the early American sea captains. Their lives were spent in a ceaseless conflict with the forces of nature and of men. They had to deal with a mutinous crew one day and with a typhoon the next. If by skillful seamanship a piratical schooner was avoided in the reaches of the Spanish Main, the resources of diplomacy would be taxed the next day to persuade some English or French colonial governor not to seize the cargo that had escaped the pirates. The captain must be a seaman, a sea-soldier, a sea-lawyer, and a sea-merchant, shut off from his principals by space which no electric current then annihilated. He must study markets, sell his cargo at the most profitable point, buy what his prophetic vision suggested would sell profitably, and sell half a dozen intermediate cargoes before returning, and even dispose of the vessel herself, if gain would result. His experience was almost as much commercial as nautical, and many of the shipping merchants who formed the aristocracy of old New York and Boston, mounted from the forecastle to the cabin, thence to the counting-room.

In a paper on the maritime trade of Salem, the Rev. George Bachelor tells of the conditions of this early seafaring, the sort of men engaged in it, and the stimulus it offered to all their faculties:

"After a century of comparative quiet, the citizens of the little town were suddenly dispersed to every part of the Oriental world, and to every nook of barbarism which had a market and a shore. The borders of the commercial world received sudden enlargement, and the boundaries of the intellectual world underwent similar expansion. The reward of enterprise might be the discovery of an island in which wild pepper enough to load a ship might be had almost for the asking, or of forests where precious gems had no commercial value, or spice islands unvisited and unvexed by civilization. Every ship-master and every mariner returning on a richly loaded ship was the custodian of valuable information. In those days crews were made up of Salem boys, every one of whom expected to become an East Indian merchant. When a captain was asked at Manila how he contrived to find his way in the teeth of a northeast monsoon by mere dead reckoning, he replied that he had a crew of twelve men, any one of whom could take and work a lunar observation as well, for all practical purposes, as Sir Isaac Newton himself.

"When, in 1816, George Coggeshall coasted the Mediterranean in the 'Cleopatra's Barge,' a magnificent yacht of 197 tons, which excited the wonder even of the Genoese, the black cook, who had once sailed with Bowditch, was found to be as competent to keep a ship's reckoning as any of the officers.

"Rival merchants sometimes drove the work of preparation night and day, when virgin markets had favors to be won, and ships which set out for unknown ports were watched when they slipped their cables and sailed away by night, and dogged for months on the high seas, in the hopes of discovering a secret, well kept by the owner and crew. Every man on board was allowed a certain space for his own little venture. People in other pursuits, not excepting the owner's minister, entrusted their savings to the supercargo, and watched eagerly the result of their adventure. This great mental activity, the profuse stores of knowledge brought by every ship's crew, and distributed, together with India shawls, blue china, and unheard-of curiosities from every savage shore, gave the community a rare alertness of intellect."

The spirit in which young fellows, scarcely attained to years of maturity, met and overcame the dangers of the deep is vividly depicted in Captain George Coggeshall's narrative of his first face-to-face encounter with death. He was in the schooner "Industry," off the Island of Teneriffe, during a heavy gale.

"Captain K. told me I had better go below, and that he would keep an outlook and take a little tea biscuit on deck. I had entered the cabin, when I felt a terrible shock. I ran to the companion-way, when I saw a ship athwart our bows. At that moment our foremast went by the board, carrying with it our main topmast. In an instant the two vessels separated, and we were left a perfect wreck. The ship showed a light for a few moments and then disappeared, leaving us to our fate. When we came to examine our situation, we found our bowsprit gone close to the knight-heads." An investigation showed that the collision had left the "Industry" in a grievous state, while the gale, ever increasing, blew directly on shore. But the sailors fought sturdily for life. "To retard the schooner's drift, we kept the wreck of the foremast, bowsprit, sails, spars, etc., fast by the bowsprit shrouds and other ropes, so that we drifted to leeward but about two miles the hour. To secure the mainmast was now the first object. I therefore took with me one of the best of the crew, and carried the end of a rope cable with us up to the mainmast head, and clenched it round the mast, while it was badly springing. We then took the cable to the windlass and hove taut, and thus effectually secured the mast.... We were then drifting directly on shore, where the cliffs were rocky, abrupt, and almost perpendicular, and were perhaps almost 1,000 feet high. At each blast of lightning we could see the surf break, whilst we heard the awful roar of the sea dashing and breaking against the rocks and caverns of this iron-bound island.

A "PINK"

"When I went below I found the captain in the act of going to bed; and as near as I can recollect, the following dialogue took place:

"'Well, Captain K., what shall we do next? We have now about six hours to pass before daylight; and, according to my calculation, we have only about three hours more drift. Still, before that time there may, perhaps, be some favorable change.'

"He replied: 'Mr. C., we have done all we can, and can do nothing more. I am resigned to my fate, and think nothing can save us.'

"I replied: 'Perhaps you are right; still, I am resolved to struggle to the last. I am too young to die; I am only twenty-one years of age, and have a widowed mother, three brothers, and a sister looking to me for support and sympathy. No, sir, I will struggle and persevere to the last.'

"'Ah,' said he, 'what can you do? Our boat will not live five minutes in the surf, and you have no other resource.'

"'I will take the boat,' said I, 'and when she fills I will cling to a spar. I will not die until my strength is exhausted and I can breathe no longer.' Here the conversation ended, when the captain covered his head with a blanket. I then wrote the substance of our misfortune in the log-book, and also a letter to my mother; rolled them up in a piece of tarred canvas; and, assisted by the carpenter, put the package into a tight keg, thinking that this might probably be thrown on shore, and thus our friends might perhaps know of our end."

Men who face Death thus sturdily are apt to overcome him. The gale lessened, the ship was patched up, the craven captain resumed command, and in two weeks' time the "Industry" sailed, sorely battered, into Santa Cruz, to find that she had been given up as lost, and her officers and crew "were looked upon as so many men risen from the dead." Young Coggeshall lived to follow the sea until gray-haired and weather-beaten, to die in his bed at last, and to tell the story of his eighty voyages in two volumes of memoirs, now growing very rare. Before he was sixteen he had made the voyage to Cadiz—a port now moldering, but which once was one of the great portals for the commerce of the world. In his second voyage, while lying in the harbor of Gibraltar, he witnessed one of the almost every-day dangers to which American sailors of that time were exposed:

"While we were lying in this port, one morning at daylight we heard firing at a distance. I took a spy-glass, and from aloft could clearly see three gunboats engaged with a large ship. It was a fine, clear morning, with scarcely wind enough to ruffle the glass-like surface of the water. During the first hour or two of this engagement the gunboats had an immense advantage; being propelled both by sails and oars, they were enabled to choose their own position. While the ship lay becalmed and unmanageable they poured grape and canister shot into her stern and bows like hailstones. At this time the ship's crew could not bring a single gun to bear upon them, and all they could do was to use their small arms through the ports and over the rails. Fortunately for the crew, the ship had thick and high bulwarks, which protected them from the fire of the enemy, so that while they were hid and screened by the boarding cloths, they could use their small arms to great advantage. At this stage of the action, while the captain, with his speaking-trumpet under his left arm, was endeavoring to bring one of his big guns to bear on one of the gunboats, a grapeshot passed through the port and trumpet and entered his chest near his shoulder-blade. The chief mate carried him below and laid him upon a mattress on the cabin floor. For a moment it seemed to dampen the ardor of the men; but it was but for an instant. The chief mate (I think his name was Randall), a gallant young man from Nantucket, then took the command, rallied, and encouraged the men to continue the action with renewed obstinacy and vigor. At this time a lateen-rigged vessel, the largest of the three privateers, was preparing to make a desperate atempt to board the ship on the larboard quarter, and, with nearly all his men on the forecastle and long bowsprit, were ready to take the final leap.

"In order to meet and frustrate the design of the enemy, the mate of the ship had one of the quarter-deck guns loaded with grape and canister shot; he then ordered all the ports on this quarter to be shut, so that the gun could not be seen; and thus were both parties prepared when the privateer came boldly up within a few yards of the ship's lee quarter. The captain, with a threatening flourish of his sword, cried out with a loud voice, in broken English: 'Strike, you damned rascal, or I will put you all to death.' At this moment a diminutive-looking man on board the 'Louisa,' with a musket, took deliberate aim through one of the waist ports, and shot him dead. Instantly the gun was run out and discharged upon the foe with deadly effect, so that the remaining few on board the privateer, amazed and astounded, were glad to give up the conflict and get off the best way they could.

"Soon after this a breeze sprung up, so that they could work their great guns to some purpose. I never shall forget the moment when I saw the Star-Spangled Banner blow out and wave gracefully in the wind, through the smoke. I also at the same moment saw with pleasure the three gunboats sailing and rowing away toward the land to make their escape. When the ship drew near the port, all the boats from the American shipping voluntarily went to assist in bringing her to anchor. She proved to be the letter-of-marque ship 'Louisa,' of Philadelphia.

"I went with our captain on board of her, and we there learned that, with the exception of the captain, not a man had been killed or wounded. The ship was terribly cut up and crippled in her sails and rigging—lifts and braces shot away; her stern was literally riddled like a grater, and both large and small shot, in great numbers, had entered her hull and were sticking to her sides. How the officers and crew escaped unhurt is almost impossible to conceive. The poor captain was immediately taken on shore, but only survived his wound a few days. He had a public funeral, and was followed to the grave by all the Americans in Gibraltar, and very many of the officers of the garrison and inhabitants of the town.

"INSTANTLY THE GUN WAS RUN OUT AND DISCHARGED"

"The ship had a rich cargo of coffee, sugar, and India goods on board, and I believe was bound for Leghorn. The gunboats belonged to Algeciras and fought under French colors, but were probably manned by the debased of all nations. I can form no idea how many were killed or wounded on board the gunboats, but from the great number of men on board, and from the length of the action, there must have been great slaughter. Neither can I say positively how long the engagement lasted; but I should think at least from three to four hours. To the chief mate too much credit can not be given for saving the ship after the captain was shot."

This action occurred in 1800, and the assailants fought under French colors, though the United States were at peace with France. It was fought within easy eyesight of Gibraltar, and therefore in British waters; but no effort was made by the British men-of-war—always plentiful there—to maintain the neutrality of the port. For sailors to be robbed or murdered, or to fight with desperation to avert robbery and murder, was then only a commonplace of the sea. Men from the safety of the adjoining shore only looked on in calm curiosity, as nowadays men look on indifferently to see the powerful freebooter of the not less troubled business sea rob, impoverish, and perhaps drive down to untimely death others who only ask to be permitted to make their little voyages unvexed by corsairs.

From a little book of memoirs of Captain Richard J. Cleveland, the curious observer can learn what it was to belong to a seafaring family in the golden days of American shipping. His was a Salem stock. His father, in 1756, when but sixteen years old, was captured by a British press-gang in the streets of Boston, and served for years in the British navy. For this compulsory servitude he exacted full compensation in later years by building and commanding divers privateers to prey upon the commerce of England. His three sons all became sailors, taking to the water like young ducks. A characteristic note of the cosmopolitanism of the young New Englander of that day is sounded in the most matter-of-fact fashion by young Cleveland in a letter from Havre: "I can't help loving home, though I think a young man ought to be at home in any part of the globe." And at home everywhere Captain Cleveland certainly was. All his life was spent in wandering over the Seven Seas, in ships of every size, from a 25-ton cutter to a 400-ton Indiaman. In those days of navigation laws, blockades, hostile cruisers, hungry privateers, and bloodthirsty pirates, the smaller craft was often the better, for it was wiser to brave nature's moods in a cockle-shell than to attract men's notice in a great ship. Captain Cleveland's voyages from Havre to the Cape of Good Hope, in a 45-ton cutter; from Calcutta to the Isle of France, in a 25-ton sloop; and Captain Coggeshall's voyage around Cape Horn in an unseaworthy pilot-boat are typical exploits of Yankee seamanship. We see the same spirit manifested occasionally nowadays when some New Englander crosses the ocean in a dory, or circumnavigates the world alone in a 30-foot sloop. But these adventures are apt to end ignominiously in a dime museum.

A noted sailor in his time was Captain Benjamin I. Trask, master of many ships, ruler of many deeps, who died in harness in 1871, and for whom the flags on the shipping in New York Bay were set at half-mast. An appreciative writer, Mr. George W. Sheldon, in *Harper's Magazine*, tells this story to show what manner of man he was; it was on the ship "Saratoga," from Havre to New York, with a crew among whom were several recently liberated French convicts:

"The first day out the new crew were very troublesome, owing in part, doubtless, to the absence of the mate, who was ill in bed and who died after a few hours. Suddenly the second mate, son of the commander, heard his father call out, 'Take hold of the wheel,' and going forward, saw him holding a sailor at arm's length. The mutineer was soon lodged in the cockpit; but all hands—the watch below and the watch on deck—came aft as if obeying a signal, with threatening faces and clenched fists. The captain, methodical and cool, ordered his son to run a line across the deck between him and the rebellious crew, and to arm the steward and the third mate.

"'Now go forward and get to work', he said to the gang, who immediately made a demonstration to break the line. 'The first man who passes that rope,' added the captain, 'I will shoot. I am going to call you one by one; if two come at a time I will shoot both.'

"The first to come forward was a big fellow in a red shirt. He had hesitated to advance when called; but the 'I will give you one more invitation, sir,' of the captain furnished him with the requisite resolution. So large were his wrists that ordinary shackles were too small to go around them, and ankle-shackles took their place. Escorted by the second and third mates to the cabin, he was made to lie flat on his stomach, while staples were driven through the chains of his handcuffs to pin him down. After eighteen of the mutineers had been similarly treated, the captain himself withdrew to the cabin and lay on a sofa, telling the second mate to call him in an hour. The next minute he was asleep with the stapled ruffians all around him."

As the ocean routes became more clearly defined, and the limitations and character of international trade more systematized, there sprung up a new type of American ship-master. The older type—and the more romantic—was the man who took his ship from Boston or New York, not knowing how many ports he might enter nor in how many markets he might have to chaffer before his return. But in time there came to be regular trade routes, over which ships went and came with almost the regularity of the great steamships on the Atlantic ferry to-day. Early in the nineteenth century the movement of both freight and passengers between New York or Boston on this side and London and Liverpool on the other began to demand regular sailings on announced days, and so the era of the American packet-ship began. Then, too, the trade with China grew to such great proportions that some of the finest fortunes America knew in the days before the "trust magnate" and the "multimillionaire"—were founded upon it. The clipper-built ship, designed to bring home the cargoes of tea in season to catch the early market, was the outcome of this trade. Adventures were still for the old-time trading captain who wandered about from port to port with miscellaneous cargoes; but the new aristocracy of the sea trod the deck of the packets and the clippers. Their ships were built all along the New England coast; but builders on the shores of Chesapeake Bay soon began to struggle for preëminence in this style of naval architecture. Thus, even in the days of wooden ships, the center of the ship-building industry began to move toward that point where it now seems definitely located. By 1815 the name "Baltimore clipper" was taken all over the world to signify the highest type of merchant vessel that man's skill could design. It was a Baltimore ship which first, in 1785, displayed the American flag in the Canton River and brought thence the first cargo of silks and teas. Thereafter, until the decline of American shipping, the Baltimore clippers led in the Chinese trade. These clippers in model were the outcome of forty years of effort to evade hostile cruisers, privateers, and pirates on the lawless seas. To be swift, inconspicuous, quick in maneuvering, and to offer a small target to the guns of the enemy, were the fundamental considerations involved in their design. Mr. Henry Hall, who, as special agent for the United States census, made in 1880 an inquiry into the history of ship-building in the United States, says in his report:

"A permanent impression has been made upon the form and rig of American vessels by forty years of war and interference. It was during that period that the shapes and fashions that prevail to-day were substantially attained. The old high poop-decks and quarter galleries disappeared with the lateen and the lug-sails on brigs, barks, and ships; the sharp stem was permanently abandoned; the curving home of the stem above the house poles went out of vogue, and vessels became longer in proportion to beam. The round bottoms were much in use, but the tendency toward a straight rise of the floor from the keel to a point half-way to the outer width of the ship became marked and popular. Hollow water-lines fore and aft were introduced; the forefoot of the hull ceased to be cut away so much, and the swell of the sides became less marked; the bows became somewhat sharper and were often made flaring above the water, and the square sprit-sail below the bowsprit was given up. American ship-builders had not yet learned to give their vessels much sheer, however, and in a majority of them the sheer line was almost straight from stem to stern; nor had they learned to divide the topsail into an upper and lower sail, and American vessels were distinguished by their short lower mast and the immense hoist of the topsail. The broadest beam was still at two-fifths the length of the hull. Hemp rigging, with broad channels and immense tops to the masts, was still retained; but the general arrangement and cut of the head, stay, square, and spanker sails at present in fashion were reached. The schooner rig had also become thoroughly popularized, especially for small vessels requiring speed; and the fast vessels of the day were the brigs and schooners, which were made long and sharp on the floor and low in the water, with considerable rake to the masts."

Such is the technical description of the changes which years of peril and of war wrought in the model of the American sailing ship. How the vessel herself, under full sail, looked when seen through the eyes of one who was a sailor, with the education of a writer and the temperament of a poet, is well told in these lines from "Two Years Before the Mast":

"Notwithstanding all that has been said about the beauty of a ship under full sail, there are very few who have ever seen a ship literally under all her sail. A ship never has all her sail upon her except when she has a light, steady breeze very nearly, but not quite, dead aft, and so regular that it can be trusted and is likely to last for some time. Then, with all her sails, light and heavy, and studding-sails on each side alow and aloft, she is the most glorious moving object in the world. Such a sight very few, even some who have been at sea a good deal, have ever beheld; for from the deck of your own vessel you can not see her as you would a separate object.

"One night, while we were in the tropics, I went out to the end of the flying jib-boom upon some duty; and, having finished it, turned around and lay over the boom for a long time, admiring the beauty of the sight before me. Being so far out from the deck, I could look at the ship as at a separate vessel; and there rose up from the water, supported only by the small black hull, a pyramid of canvas spreading far out beyond the hull and towering up almost, as it seemed in the indistinct night, into the clouds. The sea was as still as an inland lake; the light trade-wind was gently and steadily breathing from astern; the dark-blue sky was studded with the tropical stars; there was no sound but the rippling of the water under the stem; and the sails were spread out wide and high—the two lower studding-sails stretching on either side far beyond the deck; the topmost studding-sails like wings to the topsails; the topgallant studding-sails spreading fearlessly out above them; still higher the two royal studding-sails, looking like two kites flying from the same string; and highest of all the little sky-sail, the apex of the pyramid, seeming actually to touch the stars and to be out of reach of human hand. So quiet, too, was the sea, and so steady the breeze, that if these sails had been sculptured marble they could not have been more motionless—not a ripple on the surface of the canvas; not even a quivering of the extreme edges of the sail, so perfectly were they distended by the breeze. I was so lost in the sight that I forgot the presence of the man who came out with me, until he said (for he, too, rough old man-of-war's man that he was, had been gazing at the show), half to himself, still looking at the marble sails: 'How quietly they do their work!'"

The building of packet ships began in 1814, when some semblance of peace and order appeared upon the ocean, and continued until almost the time of the Civil War, when steamships had already begun to cut away the business of the old packets, and the Confederate cruisers were not needed to complete the work. But in their day these were grand examples of marine architecture. The first of the American transatlantic lines was the Black Ball line, so called from the black sphere on the white pennant which its ships displayed. This line was founded in 1815, by Isaac Wright & Company, with four ships sailing the first of every month, and making the outward run in about twenty-three days, the homeward voyage in about forty. These records were often beaten by ships of this and other lines. From thirteen to fifteen days to Liverpool was not an unknown record, but was rare enough to cause comment.

It was in this era that the increase in the size of ships began—an increase which is still going on without any sign of check. Before the War of 1812 men circumnavigated the world in vessels that would look small now carrying brick on the Tappan Zee. The performances of our frigates in 1812 first called the attention of builders to the possibilities of the bigger ship. The early packets were ships of from 400 to 500 tons each. As business grew larger ones were built—stout ships of 900 to 1100 tons, double-decked, with a poop-deck aft and a top-gallant forecastle forward. The first three-decker was the "Guy Mannering," 1419 tons, built in 1849 by William H. Webb, of New York, who later founded the college and home for ship-builders that stands on the wooded hills north of the Harlem River. In 1841, Clark & Sewall, of Bath, Me.—an historic house—built the "Rappahannock," 179.6 feet long, with a tonnage of 1133 tons. For a time she was thought to be as much of a "white elephant" as the "Great Eastern" afterwards proved to be. People flocked to study her lines on the ways and see her launched. They said only a Rothschild could afford to own her, and indeed when she appeared in the Mississippi—being built for the cotton trade—freights to Liverpool instantly fell off. But thereafter the size of ships—both packet and clippers—steadily and rapidly increased. Glancing down the long table of ships and their records prepared for the United States census, we find such notations as these.

Ship "Flying Cloud," built 1851; tonnage 1782; 374 miles in one day; from New York to San Francisco in 89 days 18 hours; in one day she made 433-1/2 miles, but reducing this to exactly 24 hours, she made 427-1/2 miles.

Ship "Comet," built 1851; tonnage 1836; beautiful model and good ship; made 332 knots in 24 hours, and 1512 knots in 120 consecutive hours.

"Sovereign of the Seas," built 1852; tonnage 2421; ran 6,245 miles in 22 days; 436 miles in one day; for four days her average was 398 miles.

"Lightning," built 1854; tonnage 2084; ran 436 miles in 24 hours, drawing 22 feet; from England to Calcutta with troops, in 87 days, beating other sailing vessels by from 16 to 40 days; from Boston to Liverpool in 13 days 20 hours.

"James Baines," built 1854, tonnage 2515; from Boston to Liverpool in 12 days 6 hours.

Three of these ships came from the historic yards of Donald McKay, at New York, one of the most famous of American ship-builders. The figures show the steady gain in size and speed that characterized the work of American ship-builders in those days. Then the United States was in truth a maritime nation. Every boy knew the sizes and records of the great ships, and each magnificent clipper had its eager partisans. Foreign trade was active. Merchants made great profit on cargoes from China, and speed was a prime element in the value of a ship. In 1840 the discovery of gold in California added a new demand for ocean shipping; the voyage around the Horn, already common enough for whalemen and men engaged in Asiatic trade, was taken by tens of thousands of adventurers. Then came the news of gold in Australia, and again demands were clamorous for more swift American ships. All nations of Europe were buyers at our shipyards, and our builders began seriously to consider whether the supply of timber would hold out. The yards of Maine and Massachusetts sent far afield for white oak knees and pine planking. Southern forests were drawn upon, and even the stately pines of Puget Sound were felled to make masts for a Yankee ship.

CHAPTER II.

THE TRANSITION FROM SAILS TO STEAM—THE CHANGE IN MARINE ARCHITECTURE—THE DEPOPULATION OF THE OCEAN—CHANGES IN THE SAILOR'S LOT—FROM WOOD TO STEEL—THE INVENTION OF THE STEAMBOAT—THE FATE OF FITCH—FULTON'S LONG STRUGGLES—OPPOSITION OF THE SCIENTISTS—THE "CLERMONT"—THE STEAMBOAT ON THE OCEAN—ON WESTERN RIVERS—THE TRANSATLANTIC PASSAGE—THE "SAVANNAH" MAKES THE FIRST CROSSING—ESTABLISHMENT OF BRITISH LINES—EFFORTS OF UNITED STATES SHIP-OWNERS TO COMPETE—THE FAMOUS COLLINS LINE—THE DECADENCE OF OUR MERCHANT MARINE—SIGNS OF ITS REVIVAL—OUR GREAT DOMESTIC SHIPPING INTEREST—AMERICA'S FUTURE ON THE SEA.

Even as recently as twenty years ago, the water front of a great seaport like New York, viewed from the harbor, showed a towering forest of tall and tapering masts, reaching high up above the roofs of the water-side buildings, crossed with slender spars hung with snowy canvas, and braced with a web of taut cordage. Across the street that passed the foot of the slips, reached out the great bowsprits or jibbooms, springing from fine-drawn bows where, above a keen cut-water, the figurehead—pride of the ship—nestled in confident strength. Neptune with his trident, Venus rising from the sea, admirals of every age and nationality, favorite heroes like Wellington and Andrew Jackson were carved, with varying skill, from stout oak, and set up to guide their vessels through tumultuous seas.

"THE WATER FRONT OF A GREAT SEAPORT LIKE NEW YORK"

To-day, alas, the towering masts, the trim yards, the web of cordage, the quaint figureheads, are gone or going fast. The docks, once so populous, seem deserted—not because maritime trade has fallen off, but because one steamship does the work that twenty stout clippers once were needed for. The clipper bow with figurehead and reaching jib-boom are gone, for the modern steamship has its bow bluff, its stem perpendicular, the "City of Rome" being the last great steamship to adhere to the old model. It is not improbable, however, that in this respect we shall see a return to old models, for the straight stem—an American invention, by the way—is held to be more dangerous in case of collisions. Many of the old-time sailing ships have been shorn of their towering masts, robbed of their canvas, and made into ignoble barges which, loaded with coal, are towed along by some fuming, fussing tugboat—as Samson shorn of his locks was made to bear the burdens of the Philistines. This transformation from sail to steam has robbed the ocean of much of its picturesqueness, and seafaring life of much of its charm, as well as of many of its dangers.

The greater size of vessels and their swifter trips under steam, have had the effect of depopulating the ocean, even in established trade routes. In the old days of ocean travel the meeting of a ship at sea was an event long to be remembered. The faint speck on the horizon, discernible only through the captain's glass, was hours in taking on the form of a ship. If a full-rigged ship, no handiwork of man could equal her impressiveness as she bore down before the wind, sail mounting on sail of billowing whiteness, until for the small hull cleaving the waves so swiftly, to carry all seemed nothing sort of marvelous. Always there was a hail and an interchange of names and ports; sometimes both vessels rounded to and boats passed and repassed. But now the courtesies of the sea have gone with its picturesqueness. Great ocean liners rushing through the deep, give each other as little heed as railway trains passing on parallel tracks. A twinkle of electric signals, or a fluttering of parti-colored flags, and each seeks its own horizon—the incident bounded by minutes where once it would have taken hours.

It would not be easy to say whether the sailor's lot has been lightened or not, by the substitution of steel for wood, of steam for sail. Perhaps the best evidence that the native-born American does not regard the change as wholly a blessing, is to be found in the fact that but few of them now follow the sea, and scarcely a vestige is left of the old New England seafaring population except in the fisheries—where sails are still the rule. Doubtless the explanation of this lies in the changed conditions of seafaring as a business. In the days which I have sketched in the first chapter, the boy of good habits and reasonable education who shipped before the mast, was fairly sure of prompt promotion to the quarter-deck, of a right to share in the profits of the voyage, and of finally owning his own ship. After 1860 all these conditions changed. Steamships, always costly to build, involved greater and greater investments as their size increased. Early in the history of steam navigation they became exclusively the property of corporations. Latterly the steamship lines have become adjuncts to great railway lines, and are conducted by the practiced stock manipulator—not by the veteran sea captain.

Richard J. Cleveland, a successful merchant navigator of the early days of the nineteenth century, when little more than a lad, undertook an enterprise, thus described by him in a letter from Havre:

"I have purchased a cutter-sloop of forty-three tons burden, on a credit of two years. This vessel was built at Dieppe and fitted out for a privateer; was taken by the English, and has been plying between Dover and Calais as a packet-boat. She has excellent accommodations and sails fast. I shall copper her, put her in ballast, trim with £1000 or £1500 sterling in cargo, and proceed to the Isle of France and Bourbon, where I expect to sell her, as well as the cargo, at a very handsome profit, and have no doubt of being well paid for my twelve months' work, calculating to be with you next August."

AN ARMED CUTTER

In such enterprises the young American sailors were always engaging—braving equally the perils of the deep and not less treacherous reefs and shoals of business but always struggling to become their own masters to command their own ships, and if possible, to carry their own cargoes. The youth of a nation that had fought for political independence, fought themselves for economic independence.

To men of this sort the conditions bred by the steam-carrying trade were intolerable. To-day a great steamship may well cost $2,000,000. It must have the favor of railway companies for cargoes, must possess expensive wharves at each end of its route, must have an army of agents and solicitors ever engaged upon its business. The boy who ships before the mast on one of them, is less likely to rise to the position of owner, than the switchman is to become railroad president—the latter progress has been known, but of the former I can not find a trace. So comparatively few young Americans choose the sea for their workshop in this day of steam.

If this book were the story of the merchant marine of all lands and all peoples, a chapter on the development of the steamship would be, perhaps, the most important, and certainly the most considerable part of it. But with the adoption of steam for ocean carriage began the decline of American shipping, a decline hastened by the use of iron, and then steel, for hulls. Though we credit ourselves—not without some protest from England— with the invention of the steamboat, the adaptation of the screw to the propulsion of vessels, and the invention of triple-expansion engines, yet it was England that seized upon these inventions and with them won, and long held, the commercial mastery of the seas. To-day (1902) it seems that economic conditions have so changed that the shipyards of the United States will again compete for the business of the world. We are building ships as good—perhaps better—than can be constructed anywhere else, but thus far we have not been able to build them as cheap. Accordingly our builders have been restricted to the construction of warships, coasters, and yachts. National pride has naturally demanded that all vessels for the navy be built in American shipyards, and a federal law has long restricted the trade between ports of the United States to ships built here. The lake shipping, too— prodigious in numbers and activity—is purely American. But until within a few years the American flag had almost disappeared from vessels engaged in international trade. Americans in many instances are the owners of ships flying the British flag, for the United States laws deny American registry— which is to a ship what citizenship is to a man—to vessels built abroad. While the result of this attempt to protect American shipyards has been to drive our flag from the ocean, there are indications now that our shipyards are prepared to build as cheaply as others, and that the flag will again figure on the high seas.

Popular history has ascribed to Robert Fulton the honor of building and navigating the first steamboat. Like claims to priority in many other inventions, this one is strenuously contested. Two years before Fulton's "Clermont" appeared on the Hudson, John Stevens, of Hoboken, built a steamboat propelled by a screw, the model of which is still in the Stevens Polytechnic Institute. Earlier still, John Fitch, of Pennsylvania, had made a steamboat, and urged it upon Franklin, upon Washington, and upon the American Philosophical Society without success; tried it then with the Spanish minister, and was offered a subsidy by the King of Spain for the exclusive right to the invention. Being a patriotic American, Fitch refused. "My invention must be first for my own country and then for all the world," said he. But later, after failing to reap any profit from his discover and finding himself deprived even of the honor of first invention, he wrote bitterly in 1792:

"The strange ideas I had at that time of serving my country, without the least suspicion that my only reward would be contempt and opprobrious names! To refuse the offer of the Spanish nation was the act of a blockhead of which I should not be guilty again."

Indeed Fitch's fortune was hard. His invention was a work of the purest originality. He was unread, uneducated, and had never so much as heard of a steam-engine when the idea of propelling boats by steam came to him. After repeated rebuffs—the lot of every inventor—he at length secured from the State of New Jersey the right to navigate its waters for a term of years. With this a stock company was formed and the first boat built and rebuilt. At first it was propelled by a single paddle at the stem; then by a series of paddles attached to an endless chain on each side of the boat; afterwards by paddle-wheels, and finally by upright oars at the side. The first test made on the Delaware River in August, 1787—twenty years before Fulton—in the presence of many distinguished citizens, some of them members of the Federal Convention, which had adjourned for the purpose, was completely successful. The boiler burst before the afternoon was over, but not before the inventor had demonstrated the complete practicability of his invention.

For ten years, struggling the while against cruel poverty, John Fitch labored to perfect his steamboat, and to force it upon the public favor, but in vain. Never in the history of invention did a new device more fully meet the traditional "long-felt want." Here was a growing nation made up of a fringe of colonies strung along an extended coast. No roads were built. Dense forests blocked the way inland but were pierced by navigable streams, deep bays, and placid sounds. The steamboat was the one thing necessary to cement American unity and speed American progress; but a full quarter of a century passed after Fitch had steamed up and down the Delaware before the new system of propulsion became commercially useful. The inventor did not live to see that day, and was at least spared the pain of seeing a later pioneer get credit for a discovery he thought his own. In 1798 he died—of an overdose of morphine—leaving behind the bitter writing: "The day will come when some powerful man will get fame and riches from my invention; but nobody will ever believe that poor John Fitch can do anything worthy of attention."

In trying to make amends for the long injustice done to poor Fitch, modern history has come near to going beyond justice. It is undoubted that Fitch applied steam to the propulsion of a boat, long before Fulton, but that Fitch himself was the first inventor is not so certain. Blasco de Garay built a rude steamboat in Barcelona in 1543; in Germany one Papin built one a few years later, which bargemen destroyed lest their business be injured by it. Jonathan Hulls, of Liverpool, in 1737 built a stern-wheeler, rude engravings of which are still in existence, and Symington in 1801 built a thoroughly practical steamboat at Dundee. 'Tis a vexed question, and perhaps it is well enough to say that Fitch first scented the commercial possibilities of steam navigation, while Fulton actually developed them—the one "raised" the fox, while the other was in at the death.

To trace a great idea to the actual birth is apt to be obstructive to national pride. It is even said that the Chinese of centuries ago understood the value of the screw-propeller—for inventing which our adoptive citizen Ericsson stands in bronze on New York's Battery.

From the time of Robert Fulton, at any rate, dates the commercial usage of the steamboat. Others had done the pioneering—Fitch on the Delaware, James Rumsey on the Potomac, William Longstreet on the Savannah, Elijah Ormsley on the waters of Rhode Island, while Samuel Morey had actually traveled by steamboat from New Haven to New York. Fulton's craft was not materially better than any of these, but it happened to be launched on

——that tide in the affairs of men

Which, taken at the flood, leads on to fortune.

But the flood of that tide did not come to Fulton without long waiting and painstaking preparation. He was the son of an Irish immigrant, and born in Pennsylvania in 1765. To inventive genius he added rather unusual gifts for drawing and painting; for a time followed the calling of a painter of miniatures and went to London to study under Benjamin West, whom all America of that day thought a genius scarcely second to Raphael or Titian. He was not, like poor Fitch, doomed to the narrowest poverty and shut out from the society of the men of light and learning of the day, for we find him, after his London experience, a member of the family of Joel Barlow, then our minister to France. By this time his ambition had forsaken art for mechanics, and he was deep in plans for diving boats, submarine torpedoes, and steamboats. Through various channels he succeeded in getting his plan for moving vessels with steam, before Napoleon—then First Consul—who ordered the Minister of Marine to treat with the inventor. The Minister in due time suggested that 10,000 francs be spent on experiments to be made in the Harbor of Brest. To this Napoleon assented, and sent Fulton to the

Institute of France to be examined as to his fitness to conduct the tests. Now the Institute is the most learned body in all France. In 1860 one of its members wrote a book to prove that the earth does not revolve upon its axis, nor move about the sun. In 1878, when Edison's phonograph was being exhibited to the eminent scientists of the Institute, one rushed wrathfully down the aisle and seizing by the collar the man who manipulated the instrument, cried out, "Wretch, we are not to be made dupes of by a ventriloquist!" So it is readily understandable that after being referred to the Institute, Fulton and his project disappeared for a long time.

The learned men of the Institute of France were not alone in their incredulity. In 1803 the Philosophical Society of Rotterdam wrote to the American Philosophical Society of Philadelphia, for information concerning the development of the steam-engine in the United States. The question was referred to Benjamin H. Latrobe, the most eminent engineer in America, and his report was published approvingly in the *Transactions*. "A sort of mania," wrote Mr. Latrobe, "had indeed prevailed and not yet entirely subsided, for impelling boats by steam-engines." But his scientific hearers would at once see that there were general objections to it which could not be overcome. "These are, first, the weight of the engine and of the fuel; second, the large space it occupies; third, the tendency of its action to rack the vessel and render it leaky; fourth, the expense of maintenance; fifth, the irregularity of its motion and the motion of the water in the boiler and cistern, and of the fuel vessel in rough weather; sixth, the difficulty arising from the liability of the paddles, or oars, to break, if light, and from the weight if made strong."

But the steamboat survived this scientific indictment in six counts. Visions proved more real than scientific reasoning.

While in the shadow of the Institute's disfavor, Fulton fell in with the new minister to France, Robert R. Livingston, and the result of this acquaintance was that America gained primacy in steam navigation, and Napoleon lost the chance to get control of an invention which, by revolutionizing navigation, might have broken that British control of the sea, that in the end destroyed the Napoleonic empire. Livingston had long taken an intelligent interest in the possibilities of steam power, and had built and tested, on the Hudson, an experimental steamboat of his own. Perhaps it was this, as much as anything, which aroused the interest of Thomas Jefferson—to whom he owed his appointment as minister to France—for Jefferson was actively interested in every sort of mechanical device, and his mind was not so scientific as to be inhospitable to new, and even, revolutionary, ideas. But Livingston was not possessed by that idea which, in later years, politicians have desired us to believe especially Jeffersonian. He was no foe to monopoly. Indeed, before he had perfected his steamboat, he used his political influence to get from New York the concession of the *exclusive* right to navigate her lakes and rivers by steam. The grant was only to be effective if within one year he should produce a boat of twenty tons, moved by steam. But he failed, and in 1801 went to France, where he found Fulton. A partnership was formed, and it was largely through Livingston's money and influence that Fulton succeeded where others, earlier in the field than he, had failed. Yet even so, it was not all easy sailing for him. "When I was building my first steamboat," he said, "the project was viewed by the public either with indifference, or with contempt as a visionary scheme. My friends, indeed, were civil, but were shy. They listened with patience to my explanations, but with a settled cast of incredulity upon their countenances. I felt the full force of the lamentation of the poet—

Truths would you teach, or save a sinking land;

All fear, none aid you, and few understand.

"THE LOUD LAUGH ROSE AT MY EXPENSE"

"As I had occasion to pass daily to and from the building yard while my boat was in progress, I have often loitered unknown near the idle groups of strangers gathered in little circles and heard various inquiries as to the object of this new vehicle. The language was uniformly that of scorn, or sneer, or ridicule. The loud laugh often rose at my expense; the dry jest; the wise calculation of losses and expenditures; the dull, but endless repetition of 'the Fulton Folly.' Never did a single encouraging remark, a bright hope, or a warm wish cross my path."

The boat which Fulton was building while the wiseacres wagged their heads and prophesied disaster, was named "The Clermont." She was 130 feet long, 18 feet wide, half-decked, and provided with a mast and sail. In the undecked part were the boiler and engine, set in masonry. The wheels were fifteen feet in diameter, with buckets four feet wide, dipping two feet into the water.

It was 1806 when Fulton came home to begin her construction. Since his luckless experience with the French Institute he had tested a steamer on the Seine; failed to interest Napoleon; tried, without success, to get the British Government to adopt his torpedo; tried and failed again with the American Government at Washington. Fulton's thoughts seemed to have been riveted on his torpedo; but Livingston was confident of the future of the steamboat, and had had an engine built for it in England, which Fulton found lying on a wharf, freight unpaid, on his return from Europe. The State of New York had meantime granted the two another monopoly of steam navigation, and gave them until 1807 to prove their ability and right. The time, though brief, proved sufficient, and on the afternoon of August 7, 1807, the "Clermont" began her epoch-making voyage. The distance to Albany—150 miles—she traversed in thirty-two hours, and the end of the passenger sloop traffic on the Hudson was begun. Within a year steamboats were plying on the Raritan, the Delaware, and Lake Champlain, and the development and use of the new invention would have been more rapid than it was, save for the monopoly rights which had been granted to Livingston and Fulton. They had the sole right to navigate by steam, the waters of New York. Well and good. But suppose the stream navigated touched both New York and New Jersey. What then? Would it be seriously asserted that a steamer owned by New Jersey citizens could not land passengers at a New York port?

Fulton and Livingston strove to protect their monopoly, and the two States were brought to the brink of war. In the end the courts settled the difficulty by establishing the exclusive control of navigable waters by the Federal Government.

From the day the "Clermont" breasted the tide of the Hudson there was no check in the conquest of the waters by steam. Up the narrowest rivers, across the most tempestuous bays, along the placid waters of Long Island Sound, coasting along the front yard of the nation from Portland to Savannah the steamboats made their way, tying the young nation indissolubly together. Curiously enough it was Livingston's monopoly that gave the first impetus to the extension of steam navigation. A mechanic by the name of Robert L. Stevens, one of the first of a family distinguished in New York and New Jersey, built a steamboat on the Hudson. After one or two trips had proved its usefulness, the possessors of the monopoly became alarmed and began proceedings against the new rival. Driven from the waters about New York, Stevens took his boat around to Philadelphia. Thus not only did he open an entirely new field of river and inland water transportation, but the trip to Philadelphia demonstrated the entire practicability of steam for use in coastwise navigation. Thereafter the vessels multiplied rapidly on all American waters. Fulton himself set up a shipyard, in which he built steam ferries, river and coastwise steamboats. In 1809 he associated himself with Nicholas J. Roosevelt, to whom credit is due for the invention of the vertical paddle-wheel, in a partnership for the purpose of putting steamboats on the great rivers of the Mississippi Valley, and in 1811 the "New Orleans" was built and navigated by Roosevelt himself, from Pittsburg to the city at the mouth of the Mississippi. The voyage took fourteen days, and before undertaking it, he descended the two rivers in a flatboat, to familiarize himself with the channel. The biographer of Roosevelt prints an interesting letter from Fulton, in which he says, "I have no pretensions to be the inventor of the steamboat. Hundreds of others have tried it and failed." Four years after Roosevelt's voyage, the "Enterprise" made for the first time in history the voyage up the Mississippi and Ohio Rivers from New Orleans to Louisville, and from that era the great rivers may be said to have been fairly opened to that commerce, which in time became the greatest agency in the building up of the nation. The Great Lakes were next to feel the quickening influence of the new motive power, but it was left for the Canadian, John Hamilton, of Queenston, to open this new field. The progress of steam navigation on both lakes and rivers will be more fully described in the chapters devoted to that topic.

So rapidly now did the use of the steamboat increase on Long Island Sound, on the rivers, and along the coast that the newspapers began to discuss gravely the question whether the supply of fuel would long hold out. The boats used wood exclusively—coal was then but little used—and despite the vast forests which covered the face of the land the price of wood in cities rose because of their demand. Mr. McMaster, the eminent historian, discovers that in 1825 thirteen steamers plying on the Hudson burned sixteen hundred cords of wood per week. Fourteen hundred cords more were used by New York ferry boats, and each trip of a Sound steamer consumed sixty cords. The American who traverses the placid waters of Long Island Sound to-day in one of the swift and splendid steamboats of the Fall River or other Sound lines, enjoys very different accommodations from those which in the second quarter of the last century were regarded as palatial. The luxury of that day was a simple sort at best. When competition became strong, the old Fulton company, then running boats to Albany, announced as a special attraction the "safety barge." This was a craft without either sails or steam, of about two hundred tons burden, and used exclusively for passengers. It boasted a spacious dining-room, ninety feet long, a deck cabin for ladies, a reading room, a promenade deck, shaded and provided with seats. One of the regular steamers of the line towed it to Albany, and its passengers were assured freedom from the noise and vibration of machinery, as well as safety from possible boiler explosions—the latter rather a common peril of steamboating in those days.

"THE DREADNAUGHT"—NEW YORK AND LIVERPOOL PACKET

It was natural that the restless mind of the American, untrammeled by traditions and impatient of convention, should turn eagerly and early to the question of crossing the ocean by steam. When the rivers had been made busy highways for puffing steamboats; when the Great Lakes, as turbulent as the ocean, and as vast as the Mediterranean, were conquered by the new marine device; when steamships plied between New York, Philadelphia, Baltimore, Savannah, and Charleston, braving what is by far more perilous than mid-ocean, the danger of tempests on a lee shore, and the shifting sands of Hatteras, there seemed to the enterprising man no reason why the passage from New York to Liverpool might not be made by the same agency. The scientific authorities were all against it. Curiously enough, the weight of scientific authority is always against anything new. Marine architects and mathematicians proved to their own satisfaction at least that no vessel could carry enough coal to cross the Atlantic, that the coal bunkers would have to be bigger than the vessel itself, in order to hold a sufficient supply for the furnaces. It is a matter of history that an eminent British scientist was engaged in delivering a lecture on this very subject in Liverpool when the "Savannah," the first steamship to cross the ocean, steamed into the harbor. It is fair, however, to add that the "Savannah's" success did not wholly destroy the contention of the opponents of steam navigation, for she made much of the passage under sail, being fitted only with what we would call now "auxiliary steam power." This was in 1819, but so slow were the shipbuilders to progress beyond what had been done with the "Savannah," that in 1835 a highly respected British scientist said in tones of authority: "As to the project which was announced in the newspapers, of making the voyage from New York to Liverpool direct by steam, it was, he had no hesitation in saying, perfectly chimerical, and they might as well talk of making a voyage from New York or Liverpool to the moon." Nevertheless, in three years from that time transatlantic steam lines were in operation, and the doom of the grand old packets was sealed.

The American who will read history free from that national prejudice which is miscalled patriotism, can not fail to be impressed by the fact that, while as a nation we have led the world in the variety and audacity of our inventions, it is nearly always some other nation that most promptly and most thoroughly utilizes the genius of our inventors. Emphatically was this the case with the application of steam power to ocean steamships. Americans showed the way, but Englishmen set out upon it and were traveling it regularly before another American vessel followed in the wake of the "Savannah." In 1838 two English steamships crossed the Atlantic to New York, the "Sirius" and the "Great Western." That was the beginning of that great fleet of British steamers which now plies up and down the Seven Seas and finds its poet laureate in Mr. Kipling. A very small beginning it was, too. The "Sirius" was of 700 tons burden and 320 horse-power; the "Great Western" was 212 feet long, with a tonnage of 1340 and engines of 400 horse-power. The "Sirius" brought seven passengers to New York, at a time when the sailing clippers were carrying from eight hundred to a thousand immigrants, and from twenty to forty cabin passengers. To those who accompanied the ship on her maiden voyage it must have seemed to justify the doubts expressed by the mathematicians concerning the practicability of designing a steamship which could carry enough coal to drive the engines all the way across the Atlantic, for the luckless "Sirius" exhausted her four hundred and fifty tons of coal before reaching Sandy Hook, and could not have made the historic passage up New York Bay under steam, except for the liberal use of spars and barrels of resin which she had in cargo. Her voyage from Cork had occupied eighteen and a half days. The "Great Western," which arrived at the same time, made the run from Queenstown in fifteen days. That two steamships should lie at anchor in New York Bay at the same time, was enough to stir the wonder and awaken the enthusiasm of the provincial New Yorkers of that day. The newspapers published editorials on the marvel, and the editor of *The Courier and Enquirer*, the chief maritime authority of the time, hazarded a prophecy in this cautious fashion:

"What may be the ultimate fate of this excitement—whether or not the expenses of equipment and fuel will admit of the employment of these vessels in the ordinary packet service—we cannot pretend to form an opinion; but of the entire feasibility of the passage of the Atlantic by steam, as far as regards safety, comfort, and dispatch, even in the roughest and most boisterous weather, the most skeptical must now cease to doubt."

Unfortunately for our national pride, the story of the development of the ocean steamship industry from this small beginning to its present prodigious proportions, is one in which we of the United States fill but a little space. We have, it is true, furnished the rich cargoes of grain, of cotton, and of cattle, that have made the ocean passage in one direction profitable for shipowners. We found homes for the millions of immigrants who crowded the "'tween decks" of steamers of every flag and impelled the companies to build bigger and bigger craft to carry the ever increasing throngs. And in these later days of luxury and wealth unparalleled, we have supplied the millionaires, whose demands for quarters afloat as gorgeous as a Fifth Avenue club have resulted in the building of floating palaces. America has supported the transatlantic lines, but almost every civilized people with a seacoast has outdone us in building the ships. For a time, indeed, it seemed that we should speedily overcome the lead that England immediately took in building steamships. Her entrance upon this industry was, as we have seen, in 1838. The United States took it up about ten years later. In 1847 the Ocean Steam Navigation Company was organized in this country and secured from the Government a contract to carry the mails between New York and Bremen. Two ships were built and regular trips made for a year or more; but when the Government contract expired and was not renewed, the venture was abandoned. About the same time the owners of one of the most famous packet lines, the Black Ball, tried the experiment of supplementing their sailing service with a steamship, but it proved unprofitable. Shortly after the New York and Havre Steamship Company, with two vessels and a postal subsidy of $150,000, entered the field and continued operations with only moderate success until 1868.

The only really notable effort of Americans in the early days of steam navigation to get their share of transatlantic trade—indeed, I might almost say the most determined effort until the present time—was that made by the projectors of the Collins line, and it ended in disaster, in heavy financial loss, and in bitter sorrow.

E.K. Collins was a New York shipping merchant, the organizer and manager of one of the most famous of the old lines of sailing packets between that port and Liverpool—the Dramatic line, so called from the fact that its ships were named after popular actors of the day. Recognizing the fact that the sailing ship was fighting a losing fight against the new style of vessels, Mr. Collins interested a number of New York merchants in a distinctly American line of transatlantic ships. It was no easy task. Capital was not over plenty in the American city which now boasts itself the financial center of the world, while the opportunities for its investment in enterprises longer proved and less hazardous than steamships were numerous. But a Government mail subsidy of $858,000 annually promised a sound financial basis, and made the task of capitalization possible. It seems not unlikely that the vicissitudes of the line were largely the result of this subsidy, for one of its conditions was extremely onerous: namely, that the vessels making twenty-six voyages annually between New York and Liverpool, should always make the passage in better time than the British Cunard line, which was then in its eighth year. However, the Collins line met the exaction bravely. Four vessels were built, the "Atlantic," "Pacific," "Arctic," and "Baltic," and the time of the fleet for the westward passage averaged eleven days, ten hours and twenty-one minutes, while the British ships averaged twelve days, nineteen hours and twenty-six minutes—a very substantial triumph for American naval architecture. The Collins liners, furthermore, were models of comfort and even of luxury for the times. They averaged a cost of $700,000 apiece, a good share of which went toward enhancing the comfort of passengers. To our English cousins these ships were at first as much of a curiosity as our vestibuled trains were a few years since. When the "Atlantic" first reached Liverpool in 1849, the townspeople by the thousand came down to the dock to examine a ship with a barber shop, fitted with the curious American barber chairs enabling the customer to recline while being shaved. The provision of a special deck-house for smokers, was another innovation, while the saloon, sixty-seven by twenty feet, the dining saloon sixty by twenty, the rich fittings of rosewood and satinwood, marble-topped tables, expensive upholstery, and stained-glass windows, decorated with patriotic designs, were for a long time the subject of admiring comment in the English press. Old voyagers who crossed in the halcyon days of the Collins line and are still taking the "Atlantic ferry," agree in saying that the increase in actual comfort is not so great as might reasonably be expected. Much of the increased expenditures of the companies has gone into more gorgeous decoration, vastly more of course into pushing for greater speed; but even in the early days there was a lavish table, and before the days of the steamships the packets offered such private accommodations in the of roomy staterooms as can be excelled only by the "cabins de luxe" of the modern liner. Aside from the question of speed, however, it is probable that the two inventions which have added most to the passengers' comfort are the electric light and artificial refrigeration.

The Collins line charged from thirty to forty dollar a ton for freight, a charge which all the modern improvements and the increase in the size of vessels, has not materially lessened. In six years, however, the corporation was practically bankrupt. The high speed required by the Government more than offset the generous subsidy, and misfortune seemed to pursue the ships. The "Arctic" came into collision with a French steamer in 1854, and went down with two hundred and twenty-two of the two hundred and sixty-eight people on board. The "Pacific" left Liverpool June 23, 1856, and was never more heard of. Shortly thereafter the subsidy was withdrawn, and the famous line went slowly down to oblivion.

It was during the best days of the Collins line that it seemed that the United States might overtake Great Britain in the race for supremacy on the ocean. In 1851 the total British steam shipping engaged in foreign trade was 65,921 tons. The United States only began building steamships in 1848, yet by 1851 its ocean-going steamships aggregated 62,390 tons. For four years our growth continued so that in 1855 we had 115,000 tons engaged in foreign trade. Then began the retrograde movement, until in 1860—before the time of the Confederate cruisers—there were; according to an official report to the National Board of Trade, "no ocean mail steamers away from our own coasts, anywhere on the globe, under the American flag, except, perhaps, on the route between New York and Havre, where two steamships may then have been in commission, which, however, were soon afterward withdrawn. The two or three steamship companies which had been in existence in New York had either failed or abandoned the business; and the entire mail, passenger, and freight traffic between Great Britain and the United States, so far as this was carried on by steam, was controlled then (as it mainly is now) by British companies." And from this condition of decadence the merchant marine of the United States is just beginning to manifest signs of recovery.

When steam had fairly established its place as the most effective power for ocean voyages of every duration, and through every zone and clime, improvements in the methods of harnessing it, and in the form and material of the ships that it was to drive, followed fast upon each other. As in the case of the invention of the steamboat, the public has commonly lightly awarded the credit for each invention to some belated experimenter who, walking more firmly along a road which an earlier pioneer had broken, attained the goal that his predecessor had sought in vain. So we find credit given almost universally to John Ericsson, the Swedish-born American, for the invention of the screw-propeller. But as early as 1770 it was suggested by John Watt, and Stevens, the American inventor, actually gave a practical demonstration of its efficiency in 1804. Ericsson perfected it in 1836, and soon thereafter the British began building steamships with screws instead of paddle-wheels. For some reason, however, not easy now to conjecture, shipbuilders clung to the paddle-wheels for vessels making the transatlantic voyage, long after they were discarded on the shorter runs along the coasts of the British isles. It so happened, too, that the first vessel to use the screw in transatlantic voyages, was also first iron ship built. She was the "Great Britain," a ship of 3,000 tons, built for the Great Western Company at Bristol, England, and intended to eclipse any ship afloat. Her hull was well on the way to completion when her designer chanced to see the "Archimedes," the first screw steamer built, and straightway changed his plans to admit the use of the new method of propulsion So from 1842 may be dated the use of both screw propellers and iron ships. We must pass hastily over the other inventions, rapidly following each other, and all designed to make ocean travel more swift, more safe, and more comfortable, and to increase the profit of the shipowner. The compound engine, which has been so developed that in place of Fulton's seven miles an hour, our ocean steamships are driven now at a speed sometimes closely approaching twenty-five miles an hour, seems already destined to give way to the turbine form of engine which, applied thus far to torpedo-boats only, has made a record of forty-four miles an hour. Iron, which stood for a revolution in 1842, has itself given way to steel. And a new force, subtile, swift, and powerful, has found endless application in the body of the great ships, so that from stem to stern-post they are a network of electric wires, bearing messages, controlling the independent engines that swing the rudder, closing water-tight compartments at the first hint of danger, and making the darkest places of the great hulls as light as day at the throwing of a switch. During the period of this wonderful advance in marine architecture ship-building in the United States languished to the point of extinction. Yachts for millionaires who could afford to pay heavily for the pleasure of flying the Stars and Stripes, ships of 2500 to 4000 tons for the coasting trade, in which no foreign-built vessel was permitted to compete, and men-of-war—very few of them before 1890—kept a few shipyards from complete obliteration. But as an industry, ship-building, which once ranked at the head of American manufactures, had sunk to a point of insignificance.

The present moment (1902) seems to show the American shipping interest in the full tide of successful reëstablishment. In Congress and in boards of trade men are arguing for and against subsidies, for and against the policy of permitting Americans to buy ships of foreign builders if they will, and fly the American flag above them. But while these things remain subjects of discussion natural causes are taking Americans again to sea. Some buy great British ships, own and manage them, even although the laws of the United States compel the flying of a foreign flag. For example, the Atlantic Transport line is owned wholly by citizens of the United States, although at the present moment all its ships fly the British flag. Two new ships are, however, being completed for this line in American shipyards, the "Minnetonka" and "Minnewaska," of 13,401 tons each. This line, started by Americans in 1887, was the first to use the so-called bilge keels, or parallel keels along each side of the hull to prevent rolling. It now has a fleet of twenty-three vessels, with a total tonnage of about 90,000, and does a heavy passenger business despite the fact that its ships were primarily designed to carry cattle. Quite as striking an illustration of the fact that capital is international, and will be invested in ships or other enterprises which promise profit quite heedless of sentimental considerations of flags, was afforded by the purchase in 1901 of the Leyland line of British steamships by an American. Immediately following this came the consolidation of ownership, or merger, of the principal British-American lines, in one great corporation, a majority of the stock of which is held by Americans. Despite their ownership on this side of the water, these ships will still fly the British flag, and a part of the contract of merger is that a British shipyard shall for ten years build all new vessels needed by the consolidated lines this situation will persist. This suggests that the actual participation of Americans in the ocean-carrying trade of the world is not to be estimated by the frequency or infrequency with which the Stars and Stripes are to be met on the ocean. It furthermore gives some indication of the rapidity with which the American flag would reappear if the law to register only ships built in American yards were repealed.

Indeed, it would appear that the law protecting American ship-builders, while apparently effective for that purpose, has destroyed American shipping. Our ship-building industry has attained respectable and even impressive proportions; but our shipping, wherever brought into competition with foreign ships, has vanished. One transatlantic line only, in 1902 displayed the American flag, and that line enjoyed special and unusual privileges, without which it probably could not have existed. In consideration of building two ships in American yards, this line, the International Navigation Company, was permitted to transfer two foreign-built ships to American registry, and a ten years' postal contract was awarded it, which guaranteed in advance the cost of construction of all the ships it was required to build. It is a fact worth noting that, while the foreign lines have been vying with each other in the construction of faster and bigger ships each year, this one has built none since its initial construction, more than a decade ago. Ten years ago its American-built ships, the "New York" and the "Paris," were the largest ships afloat; now there are eighteen larger in commission, and many building. Besides this, there are only two American lines on the Atlantic which ply to other than coastwise ports—the Pacific Mail, which is run in connection with the Panama railway, and the Admiral line, which plies between New York and the West Indies. Indeed, the Commissioner of Navigation, in his report for 1901, said:

"For serious competition with foreign nations under the conditions now imposed upon ocean navigation, we are practically limited to our registered iron and steam steel vessels, which in all number 124, of 271,378 gross tons. Those under 1,000 gross tons are not now commercially available for oversea trade. There remains 4 steamships, each of over 10,000 gross tons; 5 of between 5,000 and 6,000 gross tons; 2 of between 4,000 and 5,000 tons; 18 between 3000 and 4000 tons; 35 between 2000 and 3000 tons, and 33 between 1000 and 2000 tons; in all 97 steamships over 1000 tons, aggregating 260,325 gross tons."

Most of these are engaged in coastwise trade. The fleet of the Hamburg-American line alone, among our many foreign rivals, aggregates 515,628 gross tons.

However, we must bear in mind that this seemingly insignificant place held by the United States merchant marine represents only the part it holds in the international carrying trade of the world. Such a country as Germany must expend all its maritime energies on international trade. It has little or no river and coastwise traffic. But the United States is a little world in itself; not so very small, and of late years growing greater. Our wide extended coasts on Atlantic, Pacific, and the Mexican Gulf, are bordered by rich States crowded with a people who produce and consume more per capita than any other race. From the oceans great navigable rivers, deep bays, and placid sounds, extend into the very heart of the country. The Great Lakes are bordered by States more populous and cities more busy and enterprising than those, which in the proudest days of Rome, and Carthage and Venice skirted the Mediterranean and the Adriatic. The traffic of all these trade highways is by legislation reserved for American ships alone. On the Great Lakes has sprung up a merchant marine rivaling that of some of the foremost maritime peoples, and conducting a traffic that puts to shame the busiest maritime highways of Europe. Long Island Sound bears on its placid bosom steamships that are the marvel of the traveling public the world over. The Hudson, the Ohio, the Mississippi, are all great arteries through which the life current of trade is ceaselessly flowing. A book might be written on the one subject of the part that river navigation has played in developing the interior States of this Union. Another could well be devoted to the history of lake navigation, which it is no overstatement to pronounce the most impressive chapter in the history of the American merchant marine. In this volume, however, but brief attention can be given to either.

The figures show how honorably our whole body of shipping compares in volume to that operated by any maritime people. Our total registered shipping engaged in the fisheries, coastwise, and lake traffic, and foreign trade numbered at the beginning of 1902, 24,057 vessels, with an aggregate tonnage of 5,524,218 tons. In domestic trade alone we had 4,582,683 tons, or an amount exceeding the total tonnage of Germany and Norway combined, or of Germany and France. Only England excelled us, but her lead, which in 1860 was inconsiderable, in 1901 was prodigious; the British flag flying over no less than 14,261,254 tons of shipping, more than three times our tonnage! It is proper to note that more than two-thirds of our registered tonnage is of wood.

THERE ARE BUILDING IN AMERICAN YARDS

I have already given reasons why, in the natural course of things, this disparity between the American and the British foreign-going merchant marine will not long continue. And indeed, as this book is writing, it is apparent that its end is near. Though shipyards have multiplied fast in the last five years of the nineteenth century, the first years of the new century found them all occupied up to the very limit of their capacity. Yards that began, like the Cramps, building United States warships and finding little other work, were soon under contract to build men-of-war for Russia and Japan. The interest of the people in the navy afforded a great stimulus to shipbuilding. It is told of one of the principal yards, that its promotor went to Washington with a bid for naval construction in his pocket, but without either a shipyard or capital wherewith to build one. He secured a contract for two ships, and capital readily interested itself in his project. When that contract is out of the way the yard will enter the business of building merchant vessels, just as several yards, which long had their only support from naval contracts, are now doing. There were built in the year ending June 30, 1901, in American yards, 112 vessels of over 1000 tons each, or a total of 311,778. Many of these were lake vessels; some were wooden ships. Of modern steel steamers, built on the seaboard, there were but sixteen. At the present moment there are building in American yards, or contracted for, almost 255,325 tons of steel steamships, to be launched within a year—or 89 vessels, more than twice the output of any year in our history, and an impressive earnest for the future. Nor is this rapid increase in the ship-building activity of the United States accompanied by any reduction in the wages of the American working men. Their high wages, of which ship-builders complain, and in which everyone else rejoices, remain high. But it has been demonstrated to the satisfaction, even of foreign observers, that the highly-paid American labor is the most effective, and in the end the cheapest. Our workingmen know how to use modern tools, to make compressed air, steam, electricity do their work at every possible point, and while the United States still ranks far below England as a ship-building center, Englishmen, Germans, and Frenchmen are coming over here to learn how we build the ships that we do build. If it has not yet been demonstrated that we can build ships as cheaply as any other nation, we are so near the point of demonstration, that it may be said to be expected momentarily. With the cheapest iron in the world, we have at least succeeded in making steel, the raw material of the modern ship, cheaper than it can be made elsewhere, and that accomplished, our primacy in the matter of ship-building is a matter of the immediate future. A picturesque illustration of this change is afforded by the fact that in 1894 the plates of the "Dirigo," the first steel square-rigged vessel built in the United States, were imported from England. In 1898 we exported to England some of the plates for the "Oceanic," the largest vessel built to that time.

Even the glory, such as it may be, of building the biggest ship of the time is now well within the grasp of the United States. At this writing, indeed, the biggest ship is the "Celtic," British built, and of 20,000 tons. But the distinction is only briefly for her, for at New London, Connecticut, two ponderous iron fabrics are rising on the ways that presently shall take form as ocean steamships of 25,000 tons each, to fly the American flag, and to ply between Seattle and China. These great ships afford new illustrations of more than one point already made in this chapter. To begin with they are, of course, not constructed for any individual owner. Time was that the farmer with land sloping down to New London would put in his spare time building a staunch schooner of 200 tons, man her with his neighbors, and engage for himself in the world's carrying trade. It is rather different now. The Northern Pacific railroad directors concluded that their railroad could not be developed to its fullest earning capacity without some way of carrying to the markets of the far East the agricultural products gathered up along its line. As the tendency of the times is toward gathering all branches of a business under one control, they determined to not rely upon independent shipowners, but to build their own vessels. That meant the immediate letting of a contract for $5,000,000 worth of ship construction, and that in turn meant that there was a profit to somebody in starting an entirely new shipyard to do the work. So, suddenly, one of the sleepiest little towns in New England, Groton, opposite New London, was turned into a ship-building port. The two great Northern Pacific ships will be launched about the time this book is published, but the yard by that time will have become a permanent addition to the ship-building enterprises of the United States. So, too, all along the Atlantic coast, we find ancient shipyards where, in the very earliest colonial days, wooden vessels were built, adapting themselves to the construction of the new steel steamships.

How wonderful is the contrast between the twentieth century, steel, triple-screw, 25,000-ton, electric-lighted, 25-knot steamship, and Winthrop's little "Blessing of the Bay," or Fulton's "Clermont," or even the ships of the Collins line—floating palaces as they were called at the time! Time has made commonplace the proportions of the "Great Eastern," the marine marvel not only of her age, but of the forty years that succeeded her breaking-up as impracticable on account of size. She was 19,000 tons, 690 feet long, and built with both paddle-wheels and a screw. The "Celtic" is 700 feet long, 20,000 tons, with twin screws. The one was too big to be commercially valuable, the other has held the record for size only for a year, being already outclassed by the Northern Pacific 25,000-ton monsters. That one was a failure, the other a success, is almost wholly due to the improvements in engines, which effect economy of space both in the engine-room and in the coal bunkers. It is, by the way, rather a curious illustration of the growing luxury of life, and of ocean travel, that the first voyage of this enormous ship was made as a yacht, carrying a party of pleasure-seekers, with not a pound of cargo, through the show places of the Mediterranean.

It will be interesting to chronicle here some of the characteristics of the most modern of ocean steamships, and to show by the use of some figures, the enormous proportions to which their business has attained. For this purpose it will be necessary to use figures drawn from the records of foreign lines, and from such vessels as the "Deutschland" and the "Celtic," although the purpose of this book is to tell the story of the American merchant marine. But the figures given will be approximately correct for the great American ships now building, while there are not at present in service any American passenger ships which are fairly representative of the twentieth century liner.

The "Celtic," for example, will carry 3,294 persons, of whom 2,859 will be passengers. That is, it could furnish comfortable accommodations, heated and lighted, with ample food for all the students in Harvard University, or the University of Michigan, or Columbia University, or all in Amherst, Dartmouth, Cornell, and Williams combined. If stood on end she would almost attain the height of the Washington monument placed on the roof of the Capitol at Washington. She has nine decks, and a few years ago, if converted into a shore edifice, might fairly have been reckoned in the "skyscraper" class. Her speed, as she was built primarily for capacity is only about seventeen knots, and to attain that she burns about 260 tons of coal a day. The "Deutschland," which holds the ocean record for speed, burns nearly 600 tons of coal a day, and with it carries through the seas only 16,000 tons as against the "Celtic's" 20,000. But she is one of the modern vessels built especially to carry passengers. In her hold, huge as it is, there is room for only about 600 tons of cargo, and she seldom carries more than one-sixth of that amount. One voyage of this great ship costs about $45,000, and even

- 55 -

at that heavy expense, she is a profit earner, so great is the volume of transatlantic travel and so ready are people to pay for speed and luxury. Her coal alone costs $5,000 a trip, and the expenses of the table, laundry, etc., equal those of the most luxurious hotel.

But will ever these great liners, these huge masses of steel, guided by electricity and sped by steam, build up anew the race of American sailors? Who shall say now? To-day they are manned by Scandinavians and officered, in the main, by the seamen of the foreign nations whose flags they float. But the American is an adaptable type. He at once attends upon changing conditions and conquers them. He turned from the sea to the railroads when that seemed to be the course of progress; he may retrace his steps now that the pendulum seems to swing the other way. And if he finds under the new regime less chance for the hardy topman, no opportunity for the shrewd trader to a hundred ports, the gates closed to the man of small capital, yet be sure he will conquer fate in some way. We have seen it in the armed branch of the seafaring profession only within a few months. When the fine old sailing frigates vanished from the seas, when the "Constitution" and the "Hartford" became as obsolete as the caravels of Columbus, when a navy officer found that electricity and steam were more serious problems in his calling than sails and rigging, and a bluejacket could be with the best in his watch without ever having learned to furl a royal, then said everybody: "The naval profession has gone to the dogs. Its romance has departed. Our ships should be manned from our boiler shops, and officered from our institutions of technology. There will be no more Decaturs, Somerses, Farraguts, Cushings." And then came on the Spanish war and the rush of the "Oregon" around Cape Horn, the cool thrust of Dewey's fleet into the locked waters of Manila Bay, the plucky fight and death of Bagley at Cardenas, the braving of death by Hobson at Santiago, and the complete destruction of Cervera's fleet by Schley showed that Americans could fight as well in steel ships as in wooden ones. Nor can we doubt that the history of the next half-century will show that the new order at sea will breed a new race of American seamen able as in the past to prove themselves masters of the deep.

CHAPTER III

An Ugly Feature of Early Seafaring—The Slave Trade and Its Promoters—Part Played by Eminent New Englanders—How the Trade Grew Up—The Pious Auspices Which Surrounded the Traffic—Slave-Stealing and Sabbath-Breaking—Conditions of the Trade—Size of the Vessels—How the Captives Were Treated—Mutinies, Man-Stealing, and Murder—The Revelations of the Abolition Society—Efforts to Break Up the Trade—An Awful Retribution—England Leads the Way—Difficulty of Enforcing the Law—America's Shame—The End of the Evil—The Last Slaver.

At the foot of Narragansett Bay, with the surges of the open ocean breaking fiercely on its eastward side, and a sheltered harbor crowded with trim pleasure craft, leading up to its rotting wharves, lies the old colonial town of Newport. A holiday place it is to-day, a spot of splendor and of wealth almost without parallel in the world. From the rugged cliffs on its seaward side great granite palaces stare, many-windowed, over the Atlantic, and velvet lawns slope down to the rocks. These are the homes of the people who, in the last fifty years, have brought new life and new riches to Newport. But down in the old town you will occasionally come across a fine old colonial mansion, still retaining some signs of its former grandeur, while scattered about the island to the north are stately old farmhouses and homesteads that show clearly enough the existence in that quiet spot of wealth and comfort for these one hundred and fifty years.

Looking upon Newport to-day, and finding it all so fair, it seems hard to believe that the foundation of all its wealth and prosperity rested upon the most cruel, the most execrable, the most inhuman traffic that ever was plied by degraded men—the traffic in slaves. Yet in the old days the trade was far from being held either cruel inhuman—indeed, vessels often set sail for the Bight of Benin to swap rum for slaves, after their owners had invoked the blessing of God upon their enterprise. Nor were its promoters held by the community to be degraded. Indeed, some of the most eminent men in the community engaged in it, and its receipts were so considerable that as early as 1729 one-half of the impost levied on slaves imported into the colony was appropriated to pave the streets of the town and build its bridges—however, we are not informed that the streets were very well paved.

It was not at Newport, however, nor even in New England that the importation of slaves first began, though for reasons which I will presently show, the bulk of the traffic in them fell ultimately to New Englanders. The first African slaves in America were landed by a Dutch vessel at Jamestown,

Virginia, in 1619. The last kidnapped Africans were brought here probably some time in the latter part of 1860—for though the traffic was prohibited in 1807, the rigorous blockade of the ports of the Confederacy during the Civil War was necessary to bring it actually to an end. The amount of human misery which that frightful traffic entailed during those 240 years almost baffles the imagination. The bloody Civil War which had, perhaps, its earliest cause in the landing of those twenty blacks at Jamestown, was scarcely more than a fitting penalty, and there was justice in the fact that it fell on North and South alike, for if the South clung longest to slavery, it was the North— even abolition New England—which had most to do with establishing it on this continent.

However, it is not with slavery, but with the slave trade we have to do. Circumstances largely forced upon the New England colonies their unsavory preëminence in this sort of commerce. To begin with, their people were as we have already seen, distinctively the seafaring folk of North America. Again, one of their earliest methods of earning a livelihood was in the fisheries, and that curiously enough, led directly to the trade in slaves. To sell the great quantities of fish they dragged up from the Banks or nearer home, foreign markets must needs be found. England and the European countries took but little of this sort of provender, and moreover England, France, Holland, and Portugal had their own fishing fleets on the Banks. The main markets for the New Englanders then were the West India Islands, the Canaries, and Madeira. There the people were accustomed to a fish diet and, indeed, were encouraged in it by the frequent fastdays of the Roman Catholic church, of which most were devout members. A voyage to the Canaries with fish was commonly prolonged to the west coast of Africa, where slaves were bought with rum. Thence the vessel would proceed to the West Indies where the slaves would be sold, a large part of the purchase price being taken in molasses, which, in its turn, was distilled into rum at home, to be used for buying more slaves—for in this traffic little of actual worth was paid for the hapless captives. Fiery rum, usually adulterated and more than ever poisonous, was all the African chiefs received for their droves of human cattle. For it they sold wives and children, made bloody war and sold their captives, kidnapped and sold their human booty.

Nothing in the history of our people shows so strikingly the progress of man toward higher ideals, toward a clearer sense of the duties of humanity and the rightful relation of the strong toward the weak, than the changed sentiment concerning the slave trade. In its most humane form the thought of that traffic to-day fills us with horror. The stories of its worst phases seem almost incredible, and we wonder that men of American blood could have been such utter brutes. But two centuries ago the foremost men of New England engaged in the trade or profited by its fruits. Peter Fanueil, who-built for Boston that historic hall which we call the Cradle of Liberty, and which in later years resounded with the anti-slavery eloquence of Garrison and Phillips, was a slave owner and an actual participant in the trade. The most "respectable" merchants of Providence and Newport were active slavers—just as some of the most respectable merchants and manufacturers of to-day make merchandise of white men, women, and children, whose slavery is none the less slavery because they are driven by the fear of starvation instead of the overseer's lash. Perhaps two hundred years from now our descendants will see the criminality of our industrial system to-day, as clearly as we see the wrong in that of our forefathers. The utmost piety was observed in setting out a slave-buying expedition. The commissions were issued "by the Grace of God," divine guidance was implored for the captain who was to swap fiery rum for stolen children, and prayers were not infrequently offered for long delayed or missing slavers. George Dowing, a Massachusetts clergyman, wrote of slavery in Barbadoes: "I believe they have bought this year no less than a thousand negroes, and the more they buie, the better able they are to buie, for in a year and a half they will earne *with God's blessing*, as much as they cost." Most of the slaves brought from the coast of Guinea in New England vessels were deported again—sent to the southern States or to the West Indies for a market. The climate and the industrial conditions of New England were alike unfavorable to the growth there of slavery, and its ports served chiefly as clearing-houses for the trade. Yet there was not even among the most enlightened and leading people of the colony any moral sentiment against slavery, and from Boston to New York slaves were held in small numbers and their prices quoted in the shipping lists and newspapers like any other merchandise. Curiously enough, the first African slaves brought to Boston were sent home again and their captors prosecuted—not wholly for stealing men, but for breaking the Sabbath. It happened in this way: A Boston ship, the "Rainbow," in 1645, making the usual voyage to Madeira with staves and salt fish, touched on the coast of Guinea for a few slaves. Her captain found the English slavers on the ground already, mightily discontented, for the trade was dull. It was still the time when there was a pretense of legality about the method of procuring the slaves; they were supposed to be malefactors convicted of crime, or at the very least, prisoners taken by some native king in war. In later years the native kings, animated by an ever-growing thirst for the white man's rum, declared war in order to secure captives, and employed decoys to lure young men into the commission of crime. These devices for keeping the man-market fully supplied had not at this time been invented, and the captains of the slavers, lying off a dangerous coast in the boiling heat of a tropical country, grew restive at the long delay. Perhaps some of the rum they had brought to trade for slaves inflamed their own blood. At any rate, dragging ashore a small cannon called significantly enough a "murderer," they attacked a village, killed many of its people, and brought off a number of blacks, two of whom fell to the lot of the captain of the "Rainbow," and were by him taken to Boston. He found no profit, however, in his piratical venture, for the story coming out, he was accused in court of "murder, man-stealing, and Sabbath-breaking," and his slaves were sent home. It was wholly as merchandise that the blacks were regarded. It is impossible to believe that the brutalities of the traffic could have been tolerated so long had the idea of the essential humanity of the Africa been grasped by those who dealt in them. Instead, they were looked upon as a superior sort of cattle, but on the long voyage across the Atlantic were treated as no cattle are treated to-day in the worst "ocean tramps" in the trade. The vessels were small, many of them half the size of the lighters that ply sluggishly up and down New York harbor. Sloops, schooners, brigantines, and scows of 40 or 50 tons burden, carrying crews of nine men including the captain and mates, were the customary craft in the early days of the eighteenth century.

In his work on "The American Slave-Trade," Mr. John R. Spears gives the dimensions of some of these puny vessels which were so heavily freighted with human woe. The first American slaver of which we have record was the "Desire," of Marblehead, 120 tons. Later vessels, however, were much smaller. The sloop, "Welcome," had a capacity of 5000 gallons of molasses. The "Fame" was 79 feet long on the keel—about a large yacht's length. In 1847, some of the captured slavers had dimensions like these: The "Felicidade" 67 tons; the "Maria" 30 tons; the "Rio Bango" 10 tons. When the trade was legal and regulated by law, the "Maria" would have been permitted to carry 45 slaves—or one and one-half to each ton register. In 1847, the trade being outlawed, no regulations were observed, and this wretched little craft imprisoned 237 negroes. But even this 10-ton slaver was not the limit. Mr. Spears finds that open rowboats, no more than 24 feet long by 7 wide, landed as many as 35 children in Brazil out of say 50 with which the voyage began. But the size of the vessels made little difference in the comfort of the slaves. Greed packed the great ones equally with the small. The blacks, stowed in rows between decks, the roof barely 3 feet 10 inches above the floor on which they lay side by side, sometimes in "spoon-fashion" with from 10 to 16 inches surface-room for each, endured months of imprisonment. Often they were so packed that the head of one slave would be between the thighs of another, and in this condition they would pass the long weeks which the Atlantic passage under sail consumed. This, too, when the legality of the slave trade was recognized, and nothing but the dictates of greed led to overcrowding. Time came when the trade was put under the ban of law and made akin to piracy. Then the need for fast vessels restricted hold room and the methods of the trade attained a degree of barbarity that can not be paralleled since the days of Nero.

"A FAVORITE TRICK OF THE FLEEING SLAVER WAS TO THROW OVER SLAVES"

Shackled together "spoon-wise," as the phrase was, they suffered and sweltered through the long middle passage, dying by scores, so that often a fifth of the cargo perished during the voyage. The stories of those who took part in the effort to suppress the traffic give some idea of its frightful cruelty.

The Rev. Pascoa Grenfell Hill, a chaplain in the British navy, once made a short voyage on a slaver which his ship, the "Cleopatra," had captured. The vessel had a full cargo, and when the capture was effected, the negroes were all brought on deck for exercise and fresh air. The poor creatures quite understood the meaning of the sudden change in their masters, and kissed the hands and clothing of their deliverers. The ship was headed for the Cape of Good Hope, where the slaves were to be liberated; but a squall coming on, all were ordered below again. "The night," enters Mr. Hill in his journal, "being intensely hot, four hundred wretched beings thus crammed into a hold twelve yards in length, seven feet in breadth, and only three and one-half feet in height, speedily began to make an effort to reissue to the open air. Being thrust back and striving the more to get out, the afterhatch was forced down upon them. Over the other hatchway, in the fore part of the vessel, a wooden grating was fastened. To this, the sole inlet for the air, the suffocating heat of the hold and, perhaps, panic from the strangeness of their situation, made them flock, and thus a great part of the space below was rendered useless. They crowded to the grating and clinging to it for air, completely barred its entrance. They strove to force their way through apertures in length fourteen inches and barely six inches in breadth, and in some instances succeeded. The cries, the heat, I may say without exaggeration, the smoke of their torment which ascended can be compared to nothing earthly. One of the Spaniards gave warning that the consequences would be 'many deaths;' this prediction was fearfully verified, for the next morning 54 crushed and mangled corpses were brought to the gangway and thrown overboard. Some were emaciated from disease, many bruised and bloody. Antoine tells me that some were found strangled; their hands still grasping each others' throats."

It is of a Brazilian slaver that this awful tale is told, but the event itself was paralleled on more than one American ship. Occasionally we encounter stories of ships destroyed by an exploding magazine, and the slaves, chained to the deck, going down with the wreck. Once a slaver went ashore off Jamaica, and the officers and crew speedily got out the boats and made for the beach, leaving the human cargo to perish. When dawn broke it was seen that the slaves had rid themselves of their fetters and were busily making rafts on which the women and children were put, while the men, plunging into the sea, swam alongside, and guided the rafts toward the shore. Now mark what the white man, the supposed representative of civilization and Christianity, did. Fearing that the negroes would exhaust the store of provisions and water that had been landed, they resolved to destroy them while still in the water. As soon as the rafts came within range, those on shore opened fire with rifles and muskets with such deadly effect that between three hundred and four hundred blacks were murdered. Only thirty-four saved themselves—and for what? A few weeks later they were sold in the slave mart at Kingston.

DEALERS WHO CAME ON BOARD WERE THEMSELVES KIDNAPPED

In the early days of the trade, the captains dealt with recognized chiefs along the coast of Guinea, who conducted marauding expeditions into the interior to kidnap slaves. Rum was the purchase price, and by skillful dilution, a competent captain was able to double the purchasing value of his cargo. The trade was not one calculated to develop the highest qualities of honor, and to swindling the captains usually added theft and murder. Any negro who came near the ship to trade, or through motives of curiosity, was promptly seized and thrust below. Dealers who came on board with kidnapped negroes were themselves kidnapped after the bargain was made. Never was there any inquiry into the title of the seller. Any slave offered was bought, though the seller had no right—even under legalized slavery—to sell.

A picturesque story was told in testimony before the English House of Commons. To a certain slaver lying off the Windward coast a girl was brought in a canoe by a well-known black trader, who took his pay and paddled off. A few moments later another canoe with two blacks came alongside and inquired for the girl. They were permitted to see her and declared she had been kidnapped; but the slaver, not at all put out by that fact, refused to give her up. Thereupon the blacks paddled swiftly off after her seller, overtook, and captured him. Presently they brought him back to the deck of the ship—an article of merchandise, where he had shortly before been a merchant.

"You won't buy me," cried the captive. "I a grand trading man! I bring you slaves."

But no scruples entered the mind of the captain of the slaver. "If they will sell you I certainly will buy you," he answered, and soon the kidnapped kidnapper was in irons and thrust below in the noisome hold with the unhappy being he had sent there. A multitude of cases of negro slave-dealers being seized in this way, after disposing of their human cattle, are recorded.

It is small wonder that torn thus from home and relatives, immured in filthy and crowded holds, ill fed, denied the two great gifts of God to man—air and water—subjected to the brutality of merciless men, and wholly ignorant of the fate in store for them, many of the slaves should kill themselves. As they had a salable value the captains employed every possible device to defeat this end—every device, that is, except kind treatment, which was beyond the comprehension of the average slaver. Sometimes the slaves would try to starve themselves to death. This the captains met by torture with the cat and thumbscrews. There is a horrible story in the testimony before the English House of Commons about a captain who actually whipped a nine-months-old child to death trying to force it to eat, and then brutally compelled the mother to throw the lacerated little body overboard. Another captain found that his captives were killing themselves, in the belief that their spirits would return to their old home. By way of meeting this superstition, he announced that all who died in this way should have their heads cut off, so that if they did return to their African homes, it would be as headless spirits. The outcome of this threat was very different from what the captain had anticipated. When a number of the slaves were brought on deck to witness the beheading of the body of one of their comrades, they seized the occasion to leap overboard and were drowned. Many sought death in this way, and as they were usually good swimmers, they actually forced themselves to drown, some persistently holding their heads under water, others raising their arms high above their heads, and in one case two who died together clung to each other so that neither could swim. Every imaginable way in which death could be sought was employed by these hopeless blacks, though, indeed, the hardships of the voyage were such as to bring it often enough unsought.

When the ship's hold was full the voyage was begun, while from the suffering blacks below, unused to seafaring under any circumstances, and desperately sick in their stifling quarters, there arose cries and moans as if the cover were taken off of purgatory. The imagination recoils from the thought of so much human wretchedness.

The publications of some of the early anti-slavery associations tell of the inhuman conditions of the trade. In an unusually commodious ship carrying over six hundred slaves, we are told that "platforms, or wide shelves, were erected between the decks, extending so far from the side toward the middle of the vessel as to be capable of containing four additional rows of slaves, by which means the perpendicular height between each tier was, after allowing for the beams and platforms, reduced to three feet, six inches, so that they could not even sit in an erect posture, besides which in the men's apartment, instead of four rows, five were stowed by putting the head of one between the thighs of another." In another ship, "In the men's apartment the space allowed to each is six feet length by sixteen inches in breadth, the boys are each allowed five feet by fourteen inches, the women five feet, ten by sixteen inches, and the girls four feet by one foot each."

"A man in his coffin has more room than one of these blacks," is the terse way in which witness after witness before the British House of Commons described the miserable condition of the slaves on shipboard.

An amazing feature of this detestable traffic is the smallness and often the unseaworthiness of the vessels in which it was carried on. Few such picayune craft now venture outside the landlocked waters of Long Island Sound, or beyond the capes of the Delaware and Chesapeake. In the early days of the eighteenth century hardy mariners put out in little craft, the size of a Hudson River brick-sloop or a harbor lighter, and made the long voyage to the Canaries and the African West Coast, withstood the perils of a prolonged anchorage on a dangerous shore, went thence heavy laden with slaves to the West Indies, and so home. To cross the Atlantic was a matter of eight or ten weeks; the whole voyage would commonly take five or six months. Nor did the vessels always make up in stanchness for their diminutive proportions. Almost any weather-beaten old hulk was thought good enough for a slaver. Captain Linsday, of Newport, who wrote home from Aumboe, said: "I should be glad I cood come rite home with my slaves, for my vessel will not last to proceed far. We can see daylight all round her bow under deck." But he was not in any unusual plight. And not only the perils of the deep had to be encountered, but other perils, some bred of man's savagery, then more freely exhibited than now, others necessary to the execrable traffic in peaceful blacks. It as a time of constant wars and the seas swarmed with French privateers alert for fat prizes. When a slaver met a privateer the battle was sure to be a bloody one for on either side fought desperate men—one party following as a trade legalized piracy and violent theft of cargoes, the other employed in the violent theft of men and women, and the incitement of murder and rapine that their cargoes might be the fuller. There would have been but scant loss to mankind in most of these conflicts had privateer and slaver both gone to the bottom. Not infrequently the slavers themselves turned pirate or privateer for the time—sometimes robbing a smaller craft of its load of slaves, sometimes actually running up the black flag and turning to piracy for a permanent calling.

In addition to the ordinary risks of shipwreck or capture the slavers encountered perils peculiar to their calling. Once in a while the slaves would mutiny, though such is the gentle and almost childlike nature of the African negro that this seldom occurred. The fear of it, however, was ever present to the captains engaged in the trade, and to guard against it the slaves—always the men and sometimes the women as well—were shackled together in pairs. Sometimes they were even fastened to the floor of the dark and stifling hold in which they were immured for months at a time. If heavy weather compelled the closing of the hatches, or if disease set in, as it too often did, the morning would find the living shackled to the dead. In brief, to guard against insurrection the captains made the conditions of life so cruel that the slaves were fairly forced to revolt. In 1759 a case of an uprising that was happily successful was recorded. The slaver "Perfect," Captain Potter, lay at anchor at Mana with one hundred slaves aboard. The mate, second mate, the boatswain, and about half the crew were sent into the interior to buy some more slaves. Noticing the reduced numbers of their jailors, the slaves determined to rise. Ridding themselves of their irons, they crowded to the deck, and, all unarmed as they were, killed the captain, the surgeon, the carpenter, the cooper, and a cabin-boy. Whereupon the remainder of the crew took to the boats and boarded a neighboring slaver, the "Spencer." The captain of this craft prudently declined to board the "Perfect," and reduce the slaves to subjection again; but he had no objection to slaughtering naked blacks at long range, so he warped his craft into position and opened fire with his guns. For about an hour this butchery was continued, and then such of the slaves as still lived, ran the schooner ashore, plundered, and burnt her.

"THE ROPE WAS PUT AROUND HIS NECK"

How such insurrections were put down was told nearly a hundred years later in an official communication to Secretary of State James Buchanan, by United States Consul George W. Gordon, the story being sworn testimony before him. The case was that of the slaver "Kentucky," which carried 530 slaves. An insurrection which broke out was speedily suppressed, but fearing lest the outbreak should be repeated, the captain determined to give the wretched captives an "object lesson" by punishing the ringleaders. This is how he did it:

"They were ironed, or chained, two together, and when they were hung, a rope was put around their necks and they were drawn up to the yard-arm clear of the sail. This did not kill them, but only choked or strangled them. They were then shot in the breast and the bodies thrown overboard. If only one of two that were ironed together was to be hung, the rope was put around his neck and he was drawn up clear of the deck, and his leg laid across the rail and chopped off to save the irons and release him from his companion, who at the same time lifted up his leg until the other was chopped off as aforesaid, and he released. The bleeding negro was then drawn up, shot in the breast and thrown overboard. The legs of about one dozen were chopped off this way.

"When the feet fell on the deck they were picked up by the crew and thrown overboard, and sometimes they shot at the body while it still hung, living, and all sorts of sport was made of the business."

Forty-six men and one woman were thus done to death: "When the woman was hung up and shot, the ball did not take effect, and she was thrown overboard living, and was seen to struggle some time in the water before she sunk;" and deponent further says, "that after this was over, they brought up and flogged about twenty men and six women. The flesh of some of them where they were flogged putrified, and came off, in some cases, six or eight inches in diameter, and in places half an inch thick."

This was in 1839, a time when Americans were very sure that for civilization, progress, humanity, and the Christian virtues, they were at least on as high a plane as the most exalted peoples of the earth.

Infectious disease was one of the grave perils with which the slavers had to reckon. The overcrowding of the slaves, the lack of exercise and fresh air, the wretched and insufficient food, all combined to make grave, general sickness an incident of almost every voyage, and actual epidemics not infrequent. This was a peril that moved even the callous captains and their crews, for scurvy or yellow-jack developing in the hold was apt to sweep the decks clear as well. A most gruesome story appears in all the books on the slave trade, of the experience of the French slaver, "Rodeur." With a cargo of 165 slaves, she was on the way to Guadaloupe in 1819, when opthalmia— a virulent disease of the eyes—appeared among the blacks. It spread rapidly, though the captain, in hopes of checking its ravages, threw thirty-six negroes into the sea alive. Finally it attacked the crew, and in a short time all save one man became totally blind. Groping in the dark, the helpless sailors made shift to handle the ropes, while the one man still having eyesight clung to the wheel. For days, in this wretched state, they made their slow way along the deep, helpless and hopeless. At last a sail was sighted. The "Rodeur's" prow is turned toward it, for there is hope, there rescue! As the stranger draws nearer, the straining eyes of the French helmsman discerns something strange and terrifying about her appearance. Her rigging is loose and slovenly, her course erratic, she seems to be idly drifting, and there is no one at the wheel. A derelict, abandoned at sea, she mocks their hopes of rescue. But she is not entirely deserted, for a faint shout comes across the narrowing strip of sea and is answered from the "Rodeur." The two vessels draw near. There can be no launching of boats by blind men, but the story of the stranger is soon told. She, too, is a slaver, a Spaniard, the "Leon," and on her, too, every soul is blind from opthalmia originating among the slaves. Not even a steersman has the "Leon." All light has gone out from her, and the "Rodeur" sheers away, leaving her to an unknown fate, for never again is she heard from. How wonderful the fate—or the Providence—that directed that upon all the broad ocean teeming with ships, engaged in honest or in criminal trade, the two that should meet must be the two on which the hand of God was laid most heavily in retribution for the suffering and the woe which white men and professed Christians were bringing to the peaceful and innocent blacks of Africa.

It will be readily understood that the special and always menacing dangers attending the slave trade made marine insurance upon that sort of cargoes exceedingly high. Twenty pounds in the hundred was the usual figure in the early days. This heavy insurance led to a new form of wholesale murder committed by the captains. The policies covered losses resulting from jettisoning, or throwing overboard the cargo; they did not insure against loss from disease. Accordingly, when a slaver found his cargo infected, he would promptly throw into the sea all the ailing negroes, while still alive, in order to save the insurance. Some of the South American states, where slaves were bought, levied an import duty upon blacks, and cases are on record of captains going over their cargo outside the harbor and throwing into the sea all who by disease or for other causes, were rendered unsalable—thus saving both duty and insurance.

In the clearer light which illumines the subject to-day, the prolonged difficulty which attended the destruction of the slave trade seems incredible. It appears that two such powerful maritime nations as Great Britain and the United States had only to decree the trade criminal and it would be abandoned. But we must remember that slaves were universally regarded as property, and an attempt to interfere with the right of their owners to carry them where they would on the high seas was denounced as an interference with property rights. We see that even to-day men are very tenacious of "property rights," and the law describes them as sacred—however immoral or repugnant to common sense and common humanity they may be. So the effort to abolish the "right" of a slaver to starve, suffocate, mutilate, torture, or murder a black man in whom he had acquired a property right by the simple process of kidnapping required more than half a century to attain complete success.

The first serious blow to the slave-trade fell in 1772, when an English court declared that any slave coming into England straightway became free. That closed all English ports to the slavers. Two years after the American colonists, then on the threshold of the revolt against Great Britain, thought to put America on a like high plane, and formally resolved that they would "not purchase any slave imported after the first day of December next; after which time, we will wholly discontinue the slave-trade, and will neither be concerned in it ourselves, nor will we hire our vessels, nor sell our commodities or manufactures to those who are concerned in it." But to this praiseworthy determination the colonists were unable to live up, and in 1776, when Jefferson proposed to put into the Declaration of Independence the charge that the British King had forced the slave-trade on the colonies, a proper sense of their own guilt made the delegates oppose it.

It was in England that the first earnest effort to break up the slave-trade began. It was under the Stars and Stripes that the slavers longest protected their murderous traffic. For a time the effort of the British humanitarians was confined to the amelioration of the conditions of the trade, prescribing space to be given each slave, prescribing surgeons, and offering bounties to be paid captains who lost less than two per cent. of their cargoes on the voyage. It is not recorded that the bounty was often claimed. On the contrary, the horrors of what was called "the middle passage" grew with the greed of the slave captains. But the revelations of inhumanity made during the parliamentary investigation were too shocking for even the indifferent and callous public sentiment of that day. Humane people saw at once that to attempt to regulate a traffic so abhorrent to every sense of humanity, was for the nation to go into partnership with murderers and manstealers, and so the demand for the absolute prohibition of the traffic gained strength from the futile attempt to regulate it. Bills for its abolition failed, now in the House of Lords, then in the House of Commons; but in 1807 a law prohibiting all participation in the trade by British ships or subjects was passed. The United States moved very slowly. Individual States under the old confederation prohibited slavery within their borders, and in some cases the slave trade; but when our forefathers came together to form that Constitution under which the nation still exists, the opposition of certain Southern States was so vigorous that the best which could be done was to authorize a tax on slaves of not more than ten dollars a head, and to provide that the traffic should not be prohibited before 1808. But there followed a series of acts which corrected the seeming failure of the constitutional convention. One prohibited American citizens "carrying on the slave trade from the United States to any foreign place or country." Another forbade the introduction of slaves into the Mississippi Territory. Others made it unlawful to carry slaves to States which prohibited the traffic, or to fit out ships for the foreign slave trade, or to serve on a slaver. The discussion caused by all these measures did much to build up a healthy public sentiment, and when 1808—the date set by the Constitution—came round, a prohibitory law was passed, and the President was authorized to use the armed vessels of the United States to give it force and effect. Notwithstanding this, however, the slave trade, though now illegal and outlawed, continued for fully half a century. Slaves were still stolen on the coast of Africa by New England sea captains, subjected to the pains and horrors of the middle passage, and smuggled into Georgia or South Carolina, to be eagerly bought by the Southern planters. A Congressman estimated that 20,000 blacks were thus smuggled into the United States annually. Lafitte's nest of pirates at Barataria was a regular slave depot; so, too, was Amelia Island, Florida. The profit on a slave smuggled into the United States amounted to $350 or $500, and the temptation was too great for men to be restrained by fear of a law, which prescribed but light penalties. It is even matter of record that a governor of Georgia resigned his office to enter the smuggling trade on a large scale. The scandal was notorious, and the rapidly growing abolition sentiment demanded that Congress so amend its laws as to make manstealers at least as subject to them as other malefactors. But Congress tried the politician's device of passing laws which would satisfy the abolitionists, the slave trader, and the slave owner as well. To-day the duty of the nation seems to have been so clear that we have scant patience with the paltering policy of Congress and the Executive that permitted half a century of profitable law-breaking. But we must remember that slaves were property, that dealing in them was immensely profitable, and that while New England wanted this profit the South wanted the blacks. Macaulay said that if any considerable financial interest could be served by denying the attraction of gravitation, there would be a very vigorous attack on that great physical truth. And so, as there were many financial interests concerned in protecting slavery, every effort to effectually abolish the trade was met by an outcry and by shrewd political opposition. The slaves were better off in the United States than at home, Congress was assured; they had the blessings of Christianity; were freed from the endless wars and perils of the African jungle. Moreover, they were needed to develop the South, while in the trade, the hardy and daring sailors were trained, who in time would make the American navy the great power of the deep. Political chicanery in Congress reinforced the clamor from without, and though act after act for the destruction of the traffic was passed, none proved to be enforcible—in each was what the politicians of a later day called a "little joker," making it ineffective. But in 1820 a law was passed declaring slave-trading piracy, and punishable with death. So Congress had done its duty at last, but it was long years before the Executive rightly enforced the law.

It is needless to go into the details of the long series of Acts of Parliament and of Congress, treaties, conventions, and naval regulations, which gradually made the outlawry of the slaver on the ocean complete. In the humane work England took the lead, sacrificing the flourishing Liverpool slave-trade with all its allied interests; sacrificing, too, the immediate prosperity of its West Indian colonies, whose plantations were tilled exclusively with slave labor, and even paying heavy cash indemnity to Spain to secure her acquiescence. Unhappily, the United States was as laggard as England was active. Indeed, a curious manifestation of national pride made the American flag the slaver's badge of immunity, for the Government stubbornly—and properly— refused to grant to British cruisers the right to search vessels under our flag, and as there were few or no American men-of-war cruising on the African coast, the slaver under the Stars and Stripes was virtually immune from capture. In 1842 a treaty with Great Britain bound us to keep a considerable squadron on that coast, and thereafter there was at least some show of American hostility to the infamous traffic.

The vitality of the traffic in the face of growing international hostility is to be explained by its increasing profits. The effect of the laws passed against it was to make slaves cheaper on the coast of Africa and dearer at the markets in America. A slave that cost $20 would bring $500 in Georgia. A ship carrying 500 would bring its owners $240,000, and there were plenty of men willing to risk the penalties of piracy for a share of such prodigious profits. Moreover, the seas swarmed then with adventurous sailors—mostly of American birth—to whom the very fact that slaving was outlawed made it more attractive. The years of European war had bred up among New Englanders a daring race of privateersmen—their vocation had long been piracy in all but name, a fact which in these later days the maritime nations recognize by trying to abolish privateering by international agreement. When the wars of the early years of the nineteenth century ended the privateersmen looked about for some seafaring enterprise which promised profit. A few became pirates, more went into the slave-trade. Men of this type were not merely willing to risk their lives in a criminal calling, but were quite as ready to fight for their property as to try to save it by flight. The slavers soon began to carry heavy guns, and with desperate crews were no mean antagonists for a man-of-war. Many of the vessels that had been built for privateers were in the trade, ready to fight a cruiser or rob a smaller slaver, as chance offered. We read of some carrying as many as twenty guns, and in that sea classic, "Tom Cringle's Log," there is a story—obviously founded on fact—of a fight between a British sloop-of-war and a slaver that gives a vivid idea of the desperation with which the outlaws could fight. But sometimes the odds were hopeless, and the slaver could not hope to escape by force of arms or by flight. Then the sternness of the law, together with a foolish rule concerning the evidence necessary to convict, resulted in the murder of the slaves, not by ones or twos, but by scores, and even hundreds, at a time. For it was the unwise ruling of the courts that actual presence of slaves on a captured ship was necessary to prove that she was engaged in the unlawful trade. Her hold might reek with the odor of the imprisoned blacks, her decks show unmistakable signs of their recent presence, leg-irons and manacles might bear dumb testimony to the purpose of her voyage, informers in the crew might even betray the captain's secret; but if the boarders from the man-of-war found no negroes on the ship, she went free. What was the natural result? When a slaver, chased by a cruiser, found that capture was certain, her cargo of slaves was thrown overboard. The cruiser in the distance might detect the frightful odor that told unmistakably of a slave-ship. Her officers might hear the screams of the unhappy blacks being flung into the sea. They might even see the bodies floating in the slaver's wake; but if, on boarding the suspected craft, they found her without a single captive, they could do nothing. This was the law for many years, and because of it thousands of slaves met a cruel death as the direct result of the effort to save them from slavery. Many stories are told of these wholesale drownings. The captain of the British cruiser "Black Joke" reports of a case in which he was pursuing two slave ships:

"When chased by the tenders both put back, made all sail up the river, and ran on shore. During the chase they were seen from our vessels to throw the slaves overboard by twos, shackled together by the ankles, and left in this manner to sink or swim as best they could. Men, women, and children were seen in great numbers struggling in the water by everyone on board the two tenders, and, dreadful to relate, upward of 150 of these wretched creatures perished in this way."

In this case, the slavers did not escape conviction, though the only penalty inflicted was the seizure of their vessels. The pursuers rescued some of the drowning negroes, who were able to testify that they had been on the suspected ship, and condemnation followed. The captain of the slaver "Brillante" took no chance of such a disaster. Caught by four cruisers in a dead calm, hidden from his enemy by the night, but with no chance of escaping before dawn, this man-stealer set about planning murder on a plan so large and with such system as perhaps has not been equaled since Caligula. First he had his heaviest anchor so swung that cutting a rope would drop it. Then the chain cable was stretched about the ship, outside the rail, and held up by light bits of rope, that would give way at any stout pull. Then the slaves—600 in all—were brought up from below, open-eyed, whispering, wondering what new act in the pitiful drama of their lives this midnight summons portended. With blows and curses the sailors ranged them along the rail and bound them to the chain cable. The anchor was cut loose, plunging into the sea it carried the cable and the shackled slaves with it to the bottom. The men on the approaching man-of-war's boats, heard a great wail of many voices, a rumble, a splash, then silence, and when they reached the ship its captain politely showed them that there were no slaves aboard, and laughed at their comments on the obvious signs of the recent presence of the blacks.

"BOUND THEM TO THE CHAIN CABLE"

A favorite trick of the slaver, fleeing from a man-of-war, was to throw over slaves a few at a time in the hope that the humanity of the pursuers would impel them to stop and rescue the struggling negroes, thus giving the slave-ship a better chance of escape. Sometimes these hapless blacks thus thrown out, as legend has it Siberian peasants sometimes throw out their children as ransom to pursuing wolves, were furnished with spars or barrels to keep them afloat until the pursuer should come up; and occasionally they were even set adrift by boat-loads. It was hard on the men of the navy to steel their hearts to the cries of these castaways as the ship sped by them; but if the great evil was to be broken up it could not be by rescuing here and there a slave, but by capturing and punishing the traders. Many officers of our navy have left on record their abhorrence of the service they were thus engaged in, but at the same time expressed their conviction that it was doing the work of humanity. They were obliged to witness such human suffering as might well move the stoutest human heart. At times they were even forced to seem as merciless to the blacks as the slave-traders themselves; but in the end their work, like the merciful cruelty of the surgeon, made for good.

When a slaver was overhauled after so swift a chase that her master had no opportunity to get rid of his damning cargo, the boarding officers saw sights that scarce Inferno itself could equal. To look into her hold, filled with naked, writhing, screaming, struggling negroes was a sight that one could see once and never forget. The effluvium that arose polluted even the fresh air of the ocean, and burdened the breeze for miles to windward. The first duty of the boarding officer was to secure the officers of the craft with their papers. Not infrequently such vessels would be provided with two captains and two sets of papers, to be used according to the nationality of the warship that might make the capture; but the men of all navies cruising on the slave coast came in time to be expert in detecting such impostures. The crew once under guard, the first task was to alleviate in some degree the sufferings of the slaves. But this was no easy task, for the overcrowded vessel could not be enlarged, and its burden could in no way be decreased in mid-ocean. Even if near the coast of Africa, the negroes could not be released by the simple process of landing them at the nearest point, for the land was filled with savage tribes, the captives were commonly from the interior, and would merely have been murdered or sold anew into slavery, had they been thus abandoned. In time the custom grew up of taking them to Liberia, the free negro state established in Africa under the protection of the United States. But it can hardly be said that much advantage resulted to the individual negroes rescued by even this method, for the Liberians were not hospitable, slave traders camped upon the borders of their state, and it was not uncommon for a freed slave to find himself in a very few weeks back again in the noisome hold of the slaver. Even under the humane care of the navy officers who were put in command of captured slavers the human cattle suffered grievously. Brought on deck at early dawn, they so crowded the ships that it was almost impossible for the sailors to perform the tasks of navigation. One officer, who was put in charge of a slaver that carried 700 slaves, writes:

"They filled the waist and gangways in a fearful jam, for there were over 700 men, women, boys, and young girls. Not even a waistcloth can be permitted among slaves on board ship, since clothing even so slight would breed disease. To ward off death, ever at work on a slave ship, I ordered that at daylight the negroes should be taken in squads of twenty or more, and given a salt-water bath by the hose-pipe of the pumps. This brought renewed life after their fearful nights on the slave deck.... No one who has never seen a slave deck can form an idea of its horrors. Imagine a deck about 20 feet wide, and perhaps 120 feet long, and 5 feet high. Imagine this to be the place of abode and sleep during long, hot, healthless nights of 720 human beings! At sundown, when they were carried below, trained slaves received the poor wretches one by one, and laying each creature on his side in the wings, packed the next against him, and the next, and the next, and so on, till like so many spoons packed away they fitted into each other a living mass. Just as they were packed so must they remain, for the pressure prevented any movement or the turning of hand or foot, until the next morning, when from their terrible night of horror they were brought on deck once more, weak and worn and sick." Then, after all had come up and been splashed with salt water from the pumps, men went below to bring up the dead. There was never a morning search of this sort that was fruitless. The stench, the suffocation, the confinement, oftentimes the violence of a neighbor, brought to every dawn its tale, of corpses, and with scant gentleness all were brought up and thrown over the side to the waiting sharks. The officer who had this experience writes also that it was thirty days after capturing the slaver before he could land his helpless charges.

No great moral evil can long continue when the attention of men has been called to it, and when their consciences, benumbed by habit, have been aroused to appreciation of the fact that it is an evil. To be sure, we, with the accumulated knowledge of our ancestors and our minds filled with a horror which their teachings instilled, sometimes think that they were slow to awaken to the enormity of some evils they tolerated. So perhaps our grandchildren may wonder that we endured, and even defended, present-day conditions, which to them will appear indefensible. And so looking back on the long continuance of the slave-trade, we wonder that it could have made so pertinacious a fight for life. We marvel, too, at the character of some of the men engaged in it in its earlier and more lawful days, forgetting that their minds had not been opened, that they regarded the negro as we regard a beeve. If in some future super-refined state men should come to abstain from all animal food, perhaps the history of the Chicago stock-yards will be as appalling as is that of the Bight of Benin to-day, and that the name of Armour should be given to a great industrial school will seem as curious as to us it is inexplicable that the founder of Fanueil Hall should have dealt in human flesh.

It is, however, a chapter in the story of the American merchant sailor upon which none will wish to linger, and yet which can not be ignored. In prosecuting the search for slaves and their markets he showed the qualities of daring, of fine seamanship, of pertinacity, which have characterized him in all his undertakings; but the brutality, the greed, the inhumanity inseparable from the slave-trade make the participation of Americans in it something not pleasant to enlarge upon. It was, as I have said, not until the days of the Civil War blockade that the traffic was wholly destroyed. As late as 1860 the yacht "Wanderer," flying the New York Yacht Club's flag, owned by a club member, and sailing under the auspices of a member of one of the foremost families of the South, made several trips, and profitable ones, as a slaver. No armed vessel thought to overhaul a trim yacht, flying a private flag, and on her first trip her officers actually entertained at dinner the officers of a British cruiser watching for slavers on the African coast. But her time came, and when in 1860 the slaver, Nathaniel Gordon, a citizen of Portland, Maine, was actually hanged as a pirate, the death-blow of the slave-trade was struck. Thereafter the end came swiftly.

CHAPTER IV

THE WHALING INDUSTRY—ITS EARLY DEVELOPMENT IN NEW ENGLAND—KNOWN TO THE ANCIENTS—SHORE WHALING—BEGINNINGS OF THE DEEP-SEA FISHERIES—THE PRIZES OF WHALING—PIETY OF ITS EARLY PROMOTERS—THE RIGHT WHALE AND THE CACHALOT—A FLURRY—SOME FIGHTING WHALES—THE "ESSEX" AND THE "ANN ALEXANDER"—TYPES OF WHALERS—DECADENCE OF THE INDUSTRY—EFFECT OF OUR NATIONAL WARS—THE EMBARGO—SOME STORIES OF WHALING LIFE.

In the old "New England Primer," on which the growing minds of Yankee infants in the early days of the eighteenth century were regaled, appears a clumsy woodcut of a spouting whale, with these lines of excellent piety but doubtful rhyme:

> Whales in the sea
>
> Their Lord obey.

It is significant of the part which the whale then played in domestic economy that his familiar bulk should be utilized to "point a moral and adorn a tale" in the most elementary of books for the instruction of children. And indeed by the time the "New England Primer" was published, with its quaint lettering and rude illustrations, the whale fishery had come to be one of the chief occupations of the seafaring men of the North Atlantic States. The pursuit of this "royal fish"—as the ancient chroniclers call him in contented ignorance of the fact that he is not a fish at all—had not, indeed, originated in New England, but had been practised by all maritime peoples of whom history has knowledge, while the researches of archeologists have shown that prehistoric peoples were accustomed to chase the gigantic cetacean for his blubber, his oil, and his bone. The American Indians, in their frail canoes, the Esquimaux, in their crank kayaks, braved the fury of this aquatic monster, whose size was to that of one of his enemies as the bulk of a battle-ship is to that of a pigmy torpedo launch. But the whale fishery in vessels fitted for cruises of moderate length had its origin in Europe, where the Basques during the Middle Ages fairly drove the animals from the Bay of Biscay, which had long swarmed with them. Not a prolific breeder, the whales soon showed the effect of Europe's eagerness for oil, whalebone and ambergris, and by the beginning of the sixteenth century the industry was on the verge of extinction. Then began that search for a sea passage to India north of the continents of Europe and America, which I have described in another chapter. The passage was not discovered, but in the icy waters great schools of right whales were found, and the chase of the "royal fish" took on new vigor. Of course there was effort on the part of one nation to acquire by violence a monopoly of this profitable business, and the Dutch, who have done much in the cause of liberty, defeated the British in a naval battle at the edge of the ice before the principle of the freedom of the fisheries was accepted. To-day science has discovered substitutes for almost all of worth that the whales once supplied, and the substitutes are in the main marked improvements on the original. But in the seventeenth and eighteenth centuries the clear whale oil for illuminating purposes, the tough and supple whalebone, the spermaceti which filled the great case in the sperm-whale's head, the precious ambergris—prized even among the early Hebrews, and chronicled in the Scriptures as a thing of great price—were prizes, in pursuit of which men braved every terror of the deep, threaded the ice-floes of the Arctic, fought against the currents about Cape Horn, and steered to every corner of the Seven Seas the small, stout brigs and barks of New England make.

The whale came to the New Englander long before the New Englanders went after him. In the earliest colonial days the carcasses of whales were frequently found stranded on the beaches of Cape Cod and Long Island. Old colonial records are full of the lawsuits growing out of these pieces of treasure-trove, the finder, the owner of the land where the gigantic carrion lay stranded, and the colony all claiming ownership, or at least shares. By 1650 all the northern colonies had begun to pursue the business of shore whaling to some extent. Crews were organized, boats kept in readiness on the beach, and whenever a whale was sighted they would put off with harpoons and lances after the huge game, which, when slain, would be towed ashore, and there cut up and tried out, to the accompaniment of a prodigious clacking of gulls and a widely diffused bad smell. This method of whaling is still followed at Amagansett and Southampton, on the shore of Long Island, though the growing scarcity of whales makes catches infrequent. In the colonial days, however, it was a source of profit assiduously cultivated by coastwise communities, and both on Long Island and Cape Cod citizens were officially enjoined to watch for whales off shore. Whales were then seen daily in New York harbor, and in 1669 one Samuel Maverick recorded in a letter that thirteen whales had been taken along the south shore during the winter, and twenty in the spring.

Little by little the boat voyages after the leviathans extended further into the sea as the industry grew and the game became scarce and shy. The people of Cape Cod were the first to begin the fishery, and earliest perfected the art of "saving" the whale—that is, of securing all of value in the carcass. But the people of the little island of Nantucket brought the industry to its highest development, and spread most widely the fame of the American whaleman. Indeed, a Nantucket whaler laden with oil was the first vessel flying the Stars and Stripes that entered a British port. It is of a sailor on this craft that a patriotic anecdote, now almost classic, is told. He was unhappily deformed, and while passing along a Liverpool street was greeted by a British tar with a blow on his "humpback" and the salutation: "Hello, Jack! What you got there?" "Bunker Hill, d——n ye!" responded the Yankee. "Think you can climb it?" Far out at sea, swept ever by the Atlantic gales, a mere sand-bank, with scant surface soil to support vegetation, this island soon proved to its settlers its unfitness to maintain an agricultural people. There is a legend that an islander, weary perhaps with the effort of trying to wrest a livelihood from the unwilling soil, looked from a hilltop at the whales tumbling and spouting in the ocean. "There," he said, "is a green pasture where our children's grandchildren will go for bread." Whether the prophecy was made or not, the event occurred, for before the Revolution the American whaling fleet numbered 360 vessels, and in the banner year of the industry, 1846, 735 ships engaged in it, the major part of the fleet hailing from Nantucket. The cruises at first were toward Greenland after the so-called right whales, a variety of the cetaceans which has an added commercial value because of the baleen, or whalebone, which hangs in great strips from the roof of its mouth to its lower jaw, forming a sort of screen or sieve by which it sifts its food out of prodigious mouthfuls of sea water. This most enormous of known living creatures feeds upon very small shell-fish, swarm in the waters it frequents. Opening wide its colossal mouth, a cavity often more than fifteen feet in length, and so deep from upper to lower jaw that the flexible sheets of whalebone, sometimes ten feet long, hang straight without touching its floor, it takes a great gulp of water. Then the cavernous jaws slowly close, expelling the water through the whalebone sieve, somewhat as a Chinese laundryman sprinkles clothes, and the small marine animals which go to feed that prodigious bulk are caught in the strainer. The right whale is from 45 to 60 feet long in its maturity, and will yield about 15 tons of oil and 1500 weight of whalebone, though individuals have been known to give double this amount.

Most of the vessels which put out of Nantucket and New Bedford, in the earliest days of the industry, after whales of this sort, were not fitted with kettles and furnaces for trying out the oil at the time of the catch, as was always the custom in the sperm-whale fishery. Their prey was near at hand, their voyages comparatively short. So the fat, dripping, reeking blubber was crammed into casks, or some cases merely thrown into the ship's hold, just as it was cut from the carcass, and so brought back weeks later to the home port—a shipload of malodorous putrefaction. Old sailors who have cruised with cargoes of cattle, of green hides, and of guano, say that nothing that ever offended the olfactories of man equals the stench of a right-whaler on her homeward voyage. Scarcely even could the slave-ships compare with it. Brought ashore, this noisome mass was boiled in huge kettles, and the resulting oil sent to lighten the night in all civilized lands. England was a good customer of the colonies, and Boston shipowners did a thriving trade with oil from New Bedford or Nantucket to London. The sloops and ketches engaged in this commerce brought back, as an old letter of directions from shipowner to skipper shows, "course wicker flasketts, Allom, Copress, drum rims, head snares, shod shovells, window-glass." The trade was conducted with the same piety that we find manifested in the direction of slave-ships and privateers. In order that the oil may fetch a good price, and the voyage be speedy, the captain is commended to God, and "That hee may please to take the Conduct of you, we pray you look carefully that hee bee worshipped dayly in yor shippe, his Sabbaths Sanctifiede, and all sinne and prophainesse let bee Surpressed." In the Revolution the fisheries suffered severely from the British cruisers, and when, after peace was declared, the whalemen began coming back from the privateers, in which they had sought service, and the wharves of Nantucket, New Bedford, and New London began again to show signs of life, the Americans were confronted by the closing of their English markets. "The whale fisheries and the Newfoundland fisheries were the nurseries of British seamen," said the British ministry to John Adams, who went to London to remonstrate. "If we let Americans bring oil to London, and sell fish to our West India colonies, the British marine will decline." For a long time, therefore, the whalers had to look elsewhere than to England for a market. Nevertheless the trade grew. New Bedford, which by the middle of the nineteenth century held three-fourths of the business, took it up with great vigor. For a time Massachusetts gave bounties to encourage the industry, but it was soon strong enough to dispense with them. By 1789 the whalers found their way to the Pacific—destined in later years to be their chief fishing-ground. In that year the total whaling tonnage of Massachusetts was 10,210, with 1611 men and an annual product of 7880 barrels sperm and 13,130 barrels whale oil. Fifteen years earlier—before the war—the figures were thrice as great.

"SENDING BOAT AND MEN FLYING INTO THE AIR"

Before this period, however, whaling had taken on a new form. Deep-sea whaling, as it was called, to distinguish it from the shore fisheries, had begun long ago. Capt. Christopher Hursey, a stout Nantucket whaleman, cruising about after right whales, ran into a stiff northwest gale and was carried far out to sea. He struck a school of sperm-whales, killed one, and brought blubber home. It was not a new discovery, for the sperm-whale or cachalot, had been known for years, but the great numbers of right whales and the ease with which they were taken, had made pursuit of this nobler game uncommon. But now the fact, growing yearly more apparent, that right whales were being driven to more inaccessible haunts, made whalers turn readily to this new prey. Moreover, the sperm-whale had in him qualities of value that made him a richer prize than his Greenland cousin. True, he lacked the useful bone. His feeding habits did not necessitate a sieve, for, as beseems a giant, he devoured stout victuals, pieces of great squids—the fabled devil-fish—as big as a man's body being found in his stomach. Such a diet develops his fighting qualities, and while the right whale usually takes the steel sullenly, and dies like an overgrown seal, the cachalot fights fiercely, now diving with such a rush that he has been known to break his jaw by the fury with which he strikes the bottom at the depth of 200 fathoms; now raising his enormous bulk in air, to fall with an all-obliterating crash upon the boat which holds his tormentors, or sending boat and men flying into the air with a furious blow of his gristly flukes, or turning on his back and crunching his assailants between his cavernous jaws. Descriptions of the dying flurry of the sperm-whale are plentiful in whaling literature, many of the best of them being in that ideal whaleman's log "The Cruise of the Cachalot," by Frank T. Bullen. I quote one of these:

"Suddenly the mate gave a howl: 'Starn all—starn all! Oh, starn!' and the oars bent like canes as we obeyed—there was an upheaval of the sea just ahead; then slowly, majestically, the vast body of our foe rose into the air. Up, up it went while my heart stood still, until the whole of that immense creature hung on high, apparently motionless, and then fell—a hundred tons of solid flesh—back into the sea. On either side of that mountainous mass the waters rose in shining towers of snowy foam, which fell in their turn, whirling and eddying around us as we tossed and fell like a chip in a whirlpool. Blinded by the flying spray, baling for very life to free the boat from the water, with which she was nearly full, it was some minutes before I was able to decide whether we were still uninjured or not. Then I saw, at a little distance, the whale lying quietly. As I looked he spouted and the vapor was red with his blood. 'Starn all!' again cried our chief, and we retreated to a considerable distance. The old warrior's practised eye had detected the coming climax of our efforts, the dying agony, or 'flurry,' of the great mammal. Turning upon his side, he began to move in a circular direction, slowly at first, then faster and faster, until he was rushing round at tremendous speed, his great head raised quite out of water at times, slashing his enormous jaws. Torrents of blood poured from his spout-hole, accompanied by hoarse bellowings, as of some gigantic bull, but really caused by the laboring breath trying to pass through the clogged air-passages. The utmost caution and rapidity of manipulation of the boat was necessary to avoid his maddened rush, but this gigantic energy was short-lived. In a few minutes he subsided slowly in death, his mighty body reclined on one side, the fin uppermost waving limply as he rolled to the swell, while the small waves broke gently over the carcass in a low, monotonous surf, intensifying the profound silence that had succeeded the tumult of our conflict with the late monarch of the deep."

"SUDDENLY THE MATE GAVE A HOWL—'STARN ALL!'"

Not infrequently the sperm-whale, breaking loose from the harpoon, would ignore the boats and make war upon his chief enemy—the ship. The history of the whale fishery is full of such occurrences. The ship "Essex," of Nantucket, was attacked and sunk by a whale, which planned its campaign of destruction as though guided by human intelligence. He was first seen at a distance of several hundred yards, coming full speed for the ship. Diving, he rose again to the surface about a ship's length away, and then surged forward on the surface, striking the vessel just forward of the fore-chains. "The ship brought up as suddenly and violently as if she had struck a rock," said the mate afterward, "and trembled for few seconds like a leaf." Then she began to settle, but not fast enough to satisfy the ire of the whale. Circling around, he doubled his speed, and bore down upon the "Essex" again. This time his head fairly stove in the bows, and the ship sank so fast that the men were barely able to provision and launch the boats. Curiously enough, the monster that had thus destroyed a stout ship paid no attention whatsoever to the little boats, which would have been like nutshells before his bulk and power. But many of the men who thus escaped only went to a fate more terrible than to have gone down with their stout ship. Adrift on a trackless sea, 1000 miles from land, in open boats, with scant provision of food or water, they faced a frightful ordeal. After twenty-eight days they found an island, but it proved a desert. After leaving it the boats became separated— one being never again heard of. In the others men died fast, and at last the living were driven by hunger actually to eat the dead. Out of the captain's boat two only were rescued; out of the mate's, three. In all twelve men were sacrificed to the whale's rage.

Mere lust for combat seemed to animate this whale, for he had not been pursued by the men of the "Essex," though perhaps in some earlier meeting with men he had felt the sting of the harpoon and the searching thrust of the lance. So great is the vitality of the cachalot that it not infrequently breaks away from its pursuers, and with two or three harpoon-heads in its body lives to a ripe, if not a placid, old age. The whale that sunk the New Bedford ship "Ann Alexander" was one of these fighting veterans. With a harpoon deep in his side he turned and deliberately ran over and sunk the boat that was fast to him; then with equal deliberation sent a second boat to the bottom. This was before noon, and occurred about six miles from the ship, which bore down as fast as could be to pick up the struggling men. The whale, apparently contented with his escape, made off. But about sunset Captain Delois, iron in hand, watching from the knight-heads of the "Ann Alexander" for other whales to repair his ill-luck, saw the redoubtable fighter not far away, swimming at about a speed of five knots. At the same time the whale spied the ship. Increasing his speed to fifteen knots, he bore down upon her, and with the full force of his more than 100 tons bulk struck her "a terrible blow about two feet from the keel and just abreast of the foremast, breaking a large hole in her bottom, through which the water poured in a rushing stream." The crew had scarce time to get out the boats, with one day's provisions, but were happily picked up by a passing vessel two days later. The whale itself met retribution five months later, when it was taken by another American ship. Two of the "Ann Alexander's" harpoons were in him, his head bore deep scars, and in it were imbedded pieces of the ill-fated ship's timbers.

Instances of the combativeness of the sperm-whale are not confined to the records of the whale fishery. Even as I write I find in a current San Francisco newspaper the story of the pilot-boat "Bonita," sunk near the Farallon Islands by a whale that attacked her out of sheer wantonness and lust for fight. The "Bonita" was lying hove-to, lazily riding the swells, when in the dark—it was 10 o'clock at night—there came a prodigious shock, that threw all standing to the deck and made the pots and pans of the cook's galley jingle like a chime out of tune. From the deck the prodigious black bulk of a whale, about eighty feet long, could be made out, lying lazily half out of water near the vessel. The timbers of the "Bonita" must have been crushed by his impact, for she began to fill, and soon sank.

In this case the disaster was probably not due to any rage or malicious intent on the part of the whale. Indeed, in the days when the ocean was more densely populated with these huge animals, collision with a whale was a well-recognized maritime peril. How many of the stout vessels against whose names on the shipping list stands the fatal word "missing," came to their ends in this way can never be known; but maritime annals are full of the reports of captains who ran "bows on" into a mysterious reef where the chart showed no obstruction, but which proved to be a whale, reddening the sea with his blood, and sending the ship—not less sorely wounded—into some neighboring port to refit.

The tools with which the business of hunting the whale is pursued are simple, even rude. Steam, it is true, has succeeded to sails, and explosives have displaced the sinewy arm of the harpooner for launching the deadly shafts; but in the main the pursuit of the monsters is conducted now as it was sixty years ago, when to command a whaler was the dearest ambition of a New England coastboy. The vessels were usually brigs or barks, occasionally schooners, ranging from 100 to 500 tons. They had a characteristic architecture, due in part to the subordination of speed to carrying capacity, and further to the specially heavy timbering about the bows to withstand the crushing of the Arctic ice-pack. The bow was scarce distinguishable from the stern by its lines, and the masts stuck up straight, without that rake, which adds so much to the trim appearance of a clipper. Three peculiarities chiefly distinguished the whalers from other ships of the same general character. At the main royal-mast head was fixed the "crow's nest"—in some vessels a heavy barrel lashed to the mast, in others merely a small platform laid on the cross-trees, with two hoops fixed to the mast above, within which the lookout could stand in safety. On the deck, amidships, stood the "try-works," brick furnaces, holding two or three great kettles, in which the blubber was reduced to odorless oil. Along each rail were heavy, clumsy wooden cranes, or davits, from which hung the whale-boats—never less than five, sometimes more, while still others were lashed to the deck, for boats were the whale's sport and playthings, and seldom was a big "fish" made fast that there was not work for the ship's carpenter.

The whale-boat, evolved from the needs of this fishery, is one of the most perfect pieces of marine architecture afloat—a true adaptation of means to an end. It is clinker-built, about 27 feet long, by 6 feet beam, with a depth of about 2 feet 6 inches; sharp at both ends and clean-sided as a mackerel. Each boat carried five oarsmen, who wielded oars of from nine to sixteen feet in length, while the mate steers with a prodigious oar ten feet long. The bow oarsman is the harpooner, but when he has made fast to the whale he goes aft and takes the mate's place at the steering oar, while the latter goes forward with the lances to deal the final murderous strokes. This curious and dangerous change of position in the boat, often with a heavy sea running, and with a 100-ton whale tugging at the tug-line seems to have grown out of nothing more sensible than the insistence of mates on recognition of their rank. But a whale-boat is not the only place where a spill is threatened because some one in power insists on doing something at once useless and dangerous.

The whale-boat also carried a stout mast, rigging two sprit sails. The mast was instantly unshipped when the whale was struck. The American boats also carried centerboards, lifting into a framework extending through the center of the craft, but the English whalemen omitted these appendages. A rudder was hung over the side, for use in emergencies. Into this boat were packed, with the utmost care and system, two line-tubs, each holding from 100 to 200 fathoms of fine manila rope, one and one-half inches round, and of a texture like yellow silk; three harpoons, wood and iron, measuring about eight feet over all, and weighing about ten pounds; three lances of the finest steel, with wooden handles, in all about eight feet long; a keg of drinking water and one of biscuits; a bucket and piggin for bailing, a small spade, knives, axes, and a shoulder bomb-gun. It can be understood easily that six men, maneuvering in so crowded a boat, with a huge whale flouncing about within a few feet, a line whizzing down the center, to be caught in which meant instant death, and the sea often running high, had need to keep their wits about them.

Harpoons and lances are kept ground to a razor edge, and, propelled by the vigorous muscles of brawny whalemen, often sunk out of sight through the papery skin and soft blubber of the whale. Beyond these primitive appliances the whale fishery never progressed very far. It is true that in later days a shoulder-gun hurled the harpoon, explosive bombs replaced the lances, the ships were in some cases fitted with auxiliary steam-power, and in a few infrequent instances steam launches were employed for whale-boats. But progress was not general. The old-fashioned whaling tubs kept the seas, while the growing scarcity of the whales and the blow to the demand for oil dealt by the discovery of petroleum, checked the development of the industry.

Now the rows of whalers rotting at New Bedford's wharves, and the somnolence of Nantucket, tell of its virtual demise.

These two towns were built upon the prosperity of the whale fishery. When it languished their fortunes sunk, never to rise to their earlier heights, though cotton-spinning came to occupy the attention of the people of New Bedford, while Nantucket found a placid prosperity in entertaining summer boarders. And even during the years when whales were plentiful, and their oil still in good demand, there came periods of interruption to the trade and poverty to its followers. The Revolution first closed the seas to American ships for seven long years, and at its close the whalers found their best market—England—still shut against them. Moreover, the high seas during the closing years of the eighteenth and the opening of the nineteenth centuries were not as to-day, when a pirate is as scarce a beast of prey as a highwayman on Hounslow Heath. The Napoleonic wars had broken down men's natural sense of order and of right, and the seas swarmed with privateers, who on occasion were ready enough to turn pirates. Many whalers fell a prey to these marauders, whose operations were rather encouraged than condemned by the European nations. Both England and France were at this period endeavoring to lure the whalemen from the United Colonies by promise of special concessions in trade, or more effective protection on the high seas than their own weakling governments could assure them. Some Nantucket whalemen were indeed enticed to the new English whaling town at Dartmouth, near Halifax, or to the French town of Dunkirk. But the effort to transplant the industry did not succeed, and the years that followed, until the fateful embargo of 1807, were a period of rapid growth for the whale fishery and increasing wealth for those who pursued it. In the form of its business organization the business of whaling was the purest form of profit-sharing we have ever seen in the United States. Everybody on the ship, from captain to cabin-boy, was a partner, vitally interested in the success of the voyage. Each had his "lay"—that is to say, his proportionate share of the proceeds of the catch. Obed Macy, in his "History of Nantucket," says: "The captain's lay is generally one-seventeenth part of all obtained; the first officer's one-twenty-eighth part; the second officer's, one-forty-fifth; the third officer's, one-sixtieth; a boat-steerer's from an eightieth to a hundred-and-twentieth, and a foremast hand's, from a hundred-and-twentieth to a hundred-and-eighty-fifth each." These proportions, of course, varied—those of the men according to the ruling wages in other branches of the merchant service; those of the officers to correspond with special qualities of efficiency. All the remainder of the catch went to the owners, who put into the enterprise the ship and outfitted her for a cruise, which usually occupied three years. Their investment was therefore a heavy one, a suitable vessel of 300-tons burden costing in the neighborhood of $22,000, and her outfit $18,000 to $20,000. Not infrequently the artisans engaged in fitting out a ship were paid by being given "lays," like the sailor. In such a case the boatmaker who built the whale-boats, the ropemaker who twisted the stout, flexible manila cord to hold the whale, the sailmaker and the cooper were all interested with the crew and the owners in the success of the voyage. It was the most practical communism that industry has ever seen, and it worked to the satisfaction of all concerned as long as the whaling trade continued profitable.

The wars in which the American people engaged during the active days of the whale fishery—the Revolution, the War of 1812, and the Civil War—were disastrous to that industry, and from the depredations committed by the Confederate cruisers in the last conflict it never fully recovered. The nature of their calling made the whalemen peculiarly vulnerable to the evils of war. Cruising in distant seas, always away from home for many months, often for years, a war might be declared and fought to a finish before they knew of it. In the disordered Napoleonic days they never could tell whether the flag floating at the peak of some armed vessel encountered at the antipodes was that of friend or foe. During both the wars with England they were the special objects of the enemy's malignant attention. From the earliest days American progress in maritime enterprise was viewed by the British with apprehension and dislike. Particularly did the growth of the cod fisheries and the chase of the whale arouse transatlantic jealousy, the value of these callings as nurseries for seamen being only too plainly apparent. Accordingly the most was made of the opportunities afforded by war for crushing the whaling industry. Whalers were chased to their favorite fishing-grounds, captured, and burned. With cynical disregard of all the rules of civilized warfare—supposing war ever to be civilized—the British gave to the captured whalers only the choice of serving in British men-of-war against their own countrymen, or re-entering the whaling trade on British ships, thus building up the British whale fishery at the expense of the American. The American response to these tactics was to abandon the business during war time. In 1775 Nantucket alone had had 150 vessels, aggregating 15,000 tons, afloat in pursuit of the whale. The trade was pushed with such daring and enterprise that Edmund Burke was moved to eulogize its followers in an eloquent speech in the British House of Commons. "Neither the perseverance of Holland," he said, "nor the activity of France, nor the dexterous and firm sagacity of English enterprise, ever carried this most perilous mode of hardy industry to the extent to which it has been pushed by this most recent people." But the eloquence of Burke could not halt the British ministry in its purpose to tax the colonies despite their protests. The Revolution followed, and the whalemen of Nantucket and New Bedford stripped their vessels, sent down yards and all running rigging, stowed the sails, tied their barks and brigs to the deserted wharves and went out of business. The trade thus rudely checked had for the year preceding the outbreak of the war handled 45,000 barrels of sperm oil, 8500 barrels of right-whale oil, and 75,000 pounds of bone.

The enforced idleness of the Revolutionary days was not easily forgotten by the whalemen, and their discontent and complainings were great when the nation was again embroiled in war with Great Britain in 1812. It can not be said that their attitude in the early days of that conflict was patriotic. They had suffered—both at the hands of France and England—wrongs which might well rouse their resentment. They had been continually impressed by England, and the warships of both nations had seized American whalers for real or alleged violations of the Orders in Council or the Ostend Manifesto; but the whalemen were more eager for peace, even with the incidental perils due to war in Europe, than for war, with its enforced idleness. When Congress ordered the embargo the whalers were at first explicitly freed from its operations; but this provision being seized upon to cover evasions of the embargo, they were ultimately included. When war was finally declared, the protests of the Nantucket people almost reached the point of threatening secession. A solemn memorial was first addressed to Congress, relating the exceedingly exposed condition of the island and its favorite calling to the perils of war, and begging that the actual declaration of war might be averted. When this had availed nothing, and the young nation had rushed into battle with a courage that must seem to us now foolhardy, the Nantucketers adopted the doubtful expedient of seeking special favor from the enemy. An appeal for immunity from the ordinary acts of war was addressed to the British Admiral Cochrane, and a special envoy was sent to the British naval officer commanding the North American station, to announce the neutrality of the island and to beg immunity from assault and pillage, and assurance that one vessel would be permitted to ply unmolested between the island and the mainland. As a result of these negotiations, Nantucket formally declared her neutrality, and by town meeting voted to accede to the British demand that her people pay no taxes for the support of the United States. In all essential things the island ceased to be a part of the United States, its people neither rendering military service nor contributing to the revenues. But their submission to the British demands did not save the whale-trade, for repeated efforts to get the whalers declared neutral and exempt from capture failed.

Half a century of peace followed, during which the whaling industry rose to its highest point; but was again on the wane when the Civil War let loose upon the remaining whalemen the Confederate cruisers, the "Shenandoah" alone burning thirty-four of them. From this last stroke the industry, enfeebled by the lessened demand for its chief product, and by the greater cost and length of voyages resulting from the growing scarcity of whales, never recovered. To-day its old-time ports are deserted by traffic. Stripped of all that had salable value, its ships rot on mud-banks or at moldering wharves. The New England boy, whose ambition half a century ago was to ship on a whaler, with a boy's lay and a straight path to the quarter-deck, now goes into a city office, or makes for the West as a miner or a railroad man. The whale bids fair to become as extinct as the dodo, and the whaleman is already as rare as the buffalo.

"ROT AT MOLDERING WHARVES"

With the extension of the fishing-grounds to the Pacific began the really great days of the whale fishery. Then, from such a port as Nantucket or New Bedford a vessel would set out, to be gone three years, carrying with her the dearest hopes and ambitions of all the inhabitants. Perhaps there would be no house without some special interest in her cruise. Tradesmen of a dozen sorts supplied stores on shares. Ambitious boys of the best families sought places before the mast, for there was then no higher goal for youthful ambition than command of a whaler. Not infrequently a captain would go direct from the marriage altar to his ship, taking a young bride off on a honeymoon of three years at sea. Of course the home conditions created by this almost universal masculine employment were curious. The whaling towns were populated by women, children, and old men. The talk of the street was of big catches and the prices of oil and bone. The conversation in the shaded parlors, where sea-shells, coral, and the trophies of Pacific cruises were the chief ornaments, was of the distant husbands and sons, the perils they braved, and when they might be expected home. The solid, square houses the whalemen built, stoutly timbered as though themselves ships, faced the ocean, and bore on their ridge-pole a railed platform called the bridge, whence the watchers could look far out to sea, scanning the horizon for the expected ship. Lucky were they if she came into the harbor without half-masted flag or other sign of disaster. The profits of the calling in its best days were great. The best New London record is that of the "Pioneer," made in an eighteen-months' cruise in 1864-5. She brought back 1391 barrels of oil and 22,650 pounds of bone, all valued at $150,060. The "Envoy," of New Bedford, after being condemned as unseaworthy, was fitted out in 1847 at a cost of $8000, and sent out on a final cruise. She found oil and bone to the value of $132,450; and reaching San Francisco in the flush times, was sold for $6000. As an offset to these records, is the legend of the Nantucket captain who appeared off the harbor's mouth after a cruise of three years. "What luck, cap'n?" asked the first to board. "Well, I got nary a barrel of oil and nary a pound of bone; but I had a *mighty good sail.*"

When the bar was crossed and the ship fairly in blue water, work began. Rudyard Kipling has a characteristic story, "How the Ship Found Herself," telling how each bolt and plate, each nut, screw-thread, brace, and rivet in one of those iron tanks we now call ships adjusts itself to its work on the first voyage. On the whaler the crew had to find itself, to readjust its relations, come to know its constituent parts, and learn the ways of its superiors. Sometimes a ship was manned by men who had grown up together and who had served often on the same craft; but as a rule the men of the forecastle were a rough and vagrant lot; capable seamen, indeed, but of the adventurous and irresponsible sort, for service before the mast on a whaler was not eagerly sought by the men of the merchant service. For a time Indians were plenty, and their fine physique and racial traits made them skillful harpooners. As they became scarce, negroes began to appear among the whalemen, with now and then a Lascar, a South Sea Islander, Portuguese, and Hawaiians. The alert New Englanders, trained to the life of the sea, seldom lingered long in the forecastle, but quickly made their way to the posts of command. There they were despots, for nowhere was the discipline more severe than on whalemen. The rule was a word and a blow—and the word was commonly a curse. The ship was out for a five-years' cruise, perhaps, and the captain knew that the safety of all depended upon unquestioning obedience to his authority. Once in a while even the cowed crew would revolt, and infrequent stories of mutiny and murder appear in the record of the whale trade. The whaler, like a man-of-war, carried a larger crew than was necessary for the work of navigation, and it was necessary to devise work to keep the men employed. As a result, the ships were kept cleaner than any others in the merchant service, even though the work of trying out the blubber was necessarily productive of smoke, soot, and grease.

As a rule the voyage to the Pacific whaling waters was round Cape Horn, though occasionally a vessel made its way to the eastward and rounded the Cape of Good Hope. Almost always the world was circumnavigated before return. In early days the Pacific whalers found their game in plenty along the coast of Chili; but in time they were forced to push further and further north until the Japan Sea and Bering Sea became the favorite fishing places.

The whale was usually first sighted by the lookout in the crow's nest. A warm-blooded animal, breathing with lungs, and not with gills, like a fish, the whale is obliged to come to the surface of the water periodically to breathe. As he does so he exhales the air from his lungs through blow-holes or spiracles at the top of his head; and this warm, moist air, coming thus from his lungs into the cool air, condenses, forming a jet of vapor looking like a fountain, though there is, in fact, no spout of water. "There she blows! B-l-o-o-o-ws! Blo-o-ows!" cries the lookout at this spectacle. All is activity at once on deck, the captain calling to the lookout for the direction and character of the "pod" or school. The sperm whale throws his spout forward at an angle, instead of perpendicularly into the air, and hence is easily distinguished from right whales at a distance. The ship is then headed toward the game, coming to about a mile away. As the whale, unless alarmed, seldom swims more than two and a half miles an hour, and usually stays below only about forty-five minutes at a time, there is little difficulty in overhauling him. Then the boats are launched, the captain and a sufficient number of men staying with the ship.

"THERE SHE BLOWS"

In approaching the whale, every effort is made to come up to him at the point of least danger. This point is determined partly by the lines of the whale's vision, partly by his methods of defense. The right whale can only see dead ahead, and his one weapon is his tail, which gigantic fin, weighing several tons and measuring sometimes twenty feet across the tips of the flukes, he swings with irresistible force and all the agility of a fencer at sword-play. He, therefore, is attacked from the side, well toward his jaws. The sperm whale, however, is dangerous at both ends. His tail, though less elastic than that of the right whale, can deal a prodigious up-and-down blow, while his gigantic jaws, well garnished with sharp teeth, and capacious gullet, that readily could gulp down a man, are his chief terrors. His eyes, too, set obliquely, enable him to command the sea at all points save dead ahead, and it is accordingly from this point that the fishermen approach him. But however stealthily they move, the opportunities for disappointment are many. Big as he is, the whale is not sluggish. In an instant he may sink bodily from sight; or, throwing his flukes high in air, "sound," to be seen no more; or, casting himself bodily on the boat, blot it out of existence; or, taking it in his jaws, carry it down with him. But supposing the whale to be oblivious of its approach, the boat comes as near as seems safe, and the harpooner, poised in the bow, his knee against the bracket that steadies him, lets fly his weapon; and, hit or miss, follows it up at once with a second bent onto the same line. Some harpooners were of such strength and skill that they could hurl their irons as far as four or five fathoms. In one famous case boats from an American and British ship were in pursuit of the same whale, the British boat on the inside. It is the law of the fishery that the whale belongs to the boat that first makes fast—and many a pretty quarrel has grown out of this rule. So in this instance—seeing the danger that his rival might win the game—the American harpooner, with a prodigious effort, darted his iron clear over the rival boat and deep into the mass of blubber.

"TAKING IT IN HIS JAWS"

What a whale will do when struck no man can tell before the event. The boat-load of puffing, perspiring men who have pulled at full speed up to the monster may suddenly find themselves confronted with a furious, vindictive, aggressive beast weighing eighty tons, and bent on grinding their boat and themselves to powder; or he may simply turn tail and run. Sometimes he sounds, going down, down, down, until all the line in the boat is exhausted, and all that other boats can bend on is gone too. Then the end is thrown over with a drag, and his reappearance awaited. Sometimes he dashes off over the surface of the water at a speed of fifteen knots an hour, towing the boat, while the crew hope that their "Nantucket sleigh-ride" will end before they lose the ship for good. But once fast, the whalemen try to pull close alongside the monster. Then the mate takes the long, keen lance and plunges it deep into the great shuddering carcass, "churning" it up and down and seeking to pierce the heart or lungs. This is the moment of danger; for, driven mad with pain, the great beast rolls and thrashes about convulsively. If the boat clings fast to his side, it is in danger of being crushed or engulfed at any moment; if it retreats, he may recover himself and be off before the death-stroke can be delivered. In later days the explosive bomb, discharged from a distance, has done away with this peril; but in the palmy days of the whale fishery the men would rush into the circle of sea lashed into foam by those mighty fins, get close to the whale, as the boxer gets under the guard of his foe, smite him with lance and razor-edged spade until his spouts ran red, and to his fury there should succeed the calm of approaching death. Then the boats, pulled off. The command was "Pipes all"; and, placidly smoking in the presence of that mighty death, the whalers awaited their ship.

Stories of "fighting whales" fill the chronicles of our old whaling ports. There was the old bull sperm encountered by Captain Huntling off the River De La Plata, which is told us in a fascinating old book, "The Nimrod of the Sea." The first boat that made fast to this tough old warrior he speedily bit in two; and while her crew were swimming away from the wreck with all possible speed, the whale thrashed away at the pieces until all were reduced to small bits. Two other boats meanwhile made fast to the furious animal. Wheeling about in the foam, reddened with his blood, he crushed them as a tiger would crunch its prey. All about him were men struggling in the water—twelve of them, the crews of the two demolished boats. Of the boats themselves nothing was left big enough to float a man. The ship was miles away. Three of the sailors climbed on the back of their enemy, clinging by the harpoons and ropes still fast to him, while the others swam away for dear life, thinking only of escaping that all-engulfing jaw or the blows of that murderous tail. Now came another boat from the ship, picked up the swimmers, and cautiously rescued those perched on the whale's back from their island of shuddering flesh. The spirit of the monster was still undaunted. Though six harpoons were sunk into his body and he was dragging 300 fathoms of line, he was still in fighting mood, crunching oars, kegs, and bits of boat for more enemies to demolish. All hands made for the ship, where Captain Hunting, quite as dogged and determined as his adversary, was preparing to renew the combat. Two spare boats were fitted for use, and again the whalemen started after their foe. He, for his part, remained on the battle-ground, amid the débris of his hunters' property, and awaited attack. Nay, more; he churned the water with his mighty tail and moved forward to meet his enemy, with ready jaw to grind them to bits. The captain at the boat-oar, or steering-oar, made a mighty effort and escaped the rush; then sent an explosive bomb into the whale's vitals as he surged past. Struck unto death, the great bull went into his flurry; but in dying he rolled over the captain's boat like an avalanche, destroying it as completely as he had the three others. So man won the battle, but at a heavy cost. The whaleman who chronicled this fight says significantly: "The captain proceeded to Buenos Ayres, as much to allow his men, who were mostly green, to run away, as for the purpose of refitting, as he knew they would be useless thereafter." It was well recognized in the whaling service that men once thoroughly "gallied," or frightened, were seldom useful again; and, indeed, most of the participants in this battle did, as the captain anticipated, desert at the first port.

Curiously enough, there did not begin to be a literature of whaling until the industry went into its decadence. The old-time whalers, leading lives of continual romance and adventure, found their calling so commonplace that they noted shipwrecks, mutinies, and disaster in the struggles of the whale baldly in their logbooks, without attempt at graphic description. It is true the piety of Nantucket did result in incorporating the whale in the local hymn-book, but with what doubtful literary success these verses from the pen of Peleg Folger—himself a whaleman—will too painfully attest:

> Thou didst, O Lord, create the mighty whale,
>
> That wondrous monster of a mighty length;
>
> Vast is his head and body, vast his tail,
>
> Beyond conception his unmeasured strength.
>
> When the surface of the sea hath broke
>
> Arising from the dark abyss below,
>
> His breath appears a lofty stream of smoke,
>
> The circling waves like glittering banks of snow.
>
> And though he furiously doth us assail,
>
> Thou dost preserve us from all dangers free;
>
> He cuts our boats in pieces with his tail,

And spills us all at once into the sea.

Stories of the whale fishery are plentiful, and of late years there has been some effort made to gather these into a kind of popular history of the industry. The following incidents are gathered from a pamphlet, published in the early days of the nineteenth century, by Thomas Nevins, a New England whaler:

"A remarkable instance of the power which the whale possesses in its tail was exhibited within my own observation in the year 1807. On the 29th of May a whale was harpooned by an officer belonging to the 'Resolution.' It descended a considerable depth, and on its reappearance evinced an uncommon degree of irritation. It made such a display of its fins and tail that few of the crew were hardy enough to approach it. The captain, observing their timidity, called a boat and himself struck a second harpoon. Another boat immediately followed, and unfortunately advanced too far. The tail was again reared into the air in a terrific attitude. The impending blow was evident. The harpooner, who was directly underneath, leaped overboard, and the next moment the threatened stroke was impressed on the center of the boat, which it buried in the water. Happily no one was injured. The harpooner who leaped overboard escaped death by the act, the tail having struck the very spot on which he stood. The effects of the blow were astonishing—the keel was broken, the gunwales and every plank excepting two were cut through, and it was evident that the boat would have been completely divided, had not the tail struck directly upon a coil of lines. The boat was rendered useless.

"The Dutch ship 'Gort-Moolen,' commanded by Cornelius Gerard Ouwekaas, with a cargo of seven fish, was anchored in Greenland, in the year 1660. The captain, perceiving a whale ahead of his ship, beckoned his attendants and threw himself into a boat. He was the first to approach the whale, and was fortunate enough to harpoon it before the arrival of the second boat, which was on the advance. Jacques Vienkes, who had the direction of it, joined his captain immediately afterward, and prepared to make a second attack on the fish when it should remount to the surface. At the moment of its ascension, the boat of Vienkes, happening, unfortunately, to be perpendicularly above it, was so suddenly and forcibly lifted up by a stroke of the head of the whale that it was dashed to pieces before the harpooner could discharge his weapon. Vienkes flew along with the pieces of the boat, and fell upon the back of the animal. This intrepid seaman, who still retained his weapon in his grasp, harpooned the whale on which he stood; and by means of the harpoon and the line, which he never abandoned, he steadied himself firmly upon the fish, notwithstanding his hazardous situation, and regardless of a considerable wound that he received in his leg in his fall along with the fragments of the boat. All the efforts of the other boats to approach the whale and deliver the harpooner were futile. The captain, not seeing any other method of saving his unfortunate companion, who was in some way entangled with the line, called him to cut it with his knife and betake himself to swimming. Vienkes, embarrassed and disconcerted as he was, tried in vain to follow this council. His knife was in the pocket of his drawers, and being unable to support himself with one hand, he could not get it out. The whale, meanwhile, continued advancing along the surface of the water with great rapidity, but fortunately never attempted to dive. While his comrades despaired of his life, the harpoon by which he held at length disengaged itself from the body of the whale. Vienkes, being thus liberated, did not fail to take advantage of this circumstance. He cast himself into the sea, and by swimming endeavored to regain the boats, which continued the pursuit of the whale. When his shipmates perceived him struggling with the waves, they redoubled their exertions. They reached him just as his strength was exhausted, and had the happiness of rescuing this adventurous harpooner from his perilous situation.

"Captain Lyons, of the 'Raith,' of Leith, while prosecuting the whale fishery on the Labrador coast, in the season of 1802, discovered a large whale at a short distance from the ship. Four boats were dispatched in pursuit, and two of them succeeded in approaching it so closely together that two harpoons were struck at the same moment. The fish descended a few fathoms in the direction of another of the boats, which was on the advance, rose accidentally beneath it, struck it with his head, and threw the boat, men, and apparatus about fifteen feet in the air. It was inverted by the stroke, and fell into the water with its keel upward. All the people were picked up alive by the fourth boat, which was just at hand, excepting one man, who, having got entangled in the boat, fell beneath it and was unfortunately drowned. The fish was soon afterward killed.

"In 1822 two boats belonging to the ship 'Baffin' went in pursuit of a whale. John Carr was harpooner and commander of them. The whale they pursued led them into a vast shoal of his own species. They were so numerous that their blowing was incessant, and they believed that they did not see fewer than a hundred. Fearful of alarming them without striking any, they remained a while motionless. At last one rose near Carr's boat, and he approached and, fatally for himself, harpooned it. When he struck, the fish was approaching the boat; and, passing very rapidly, jerked the line out of its place over the stern and threw it upon the gunwale. Its pressure in this unfavorable position so careened the boat that the side was pulled under water and it began to fill. In this emergency Carr, who was a brave, active man, seized the line, and endeavored to release the boat by restoring it to its place; but by some circumstance which was never accounted for, a turn of the line flew over his arm, dragged him overboard in an instant, and drew him under the water, never more to rise. So sudden was the accident that only one man, who was watching him, saw what had happened; so that when the boat righted, which it immediately did, though half full of water, the whole crew, on looking round, inquired what had become of Carr. It is impossible to imagine a death more awfully sudden and unexpected. The invisible bullet could not have effected more instantaneous destruction. The velocity of the whale at its first descent is from thirteen to fifteen feet per second. Now, as this unfortunate man was adjusting the line at the water's very edge, where it must have been perfectly tight, owing to its obstruction in running out of the boat, the interval between the fastening of the line about him and his disappearance could not have exceeded the third part of a second of time, for in one second only he must have been dragged ten or twelve feet deep. Indeed, he had not time for the least exclamation; and the person who saw his removal observed that it was so exceeding quick that, though his eye was upon him at the moment, he could scarcely distinguish his figure as he disappeared.

"As soon as the crew recovered from their consternation, they applied themselves to the needful attention which the lines required. A second harpoon was struck from the accompanying boat, on the rising of the whale to the surface, and some lances were applied; but this melancholy occurrence had cast such a damp on all present that they became timid and inactive in their subsequent duties. The whale, when nearly exhausted, was allowed to remain some minutes unmolested, till, having recovered some degree of energy, it made a violent effort and tore itself away from the harpoons. The exertions of the crews thus proved fruitless, and were attended with serious loss.

"A harpooner belonging to the 'Henrietta,' of Whitby, when engaged in lancing a whale into which he had previously struck a harpoon, incautiously cast a little line under his feet that he had just hauled into the boat, after it had been drawn out by the fish. A painful stroke of his lance induced the whale to dart suddenly downward. His line began to run out from under his feet, and in an instant caught him by a turn round his body. He had but just time to cry out, 'Clear away the line! Oh, dear!' when he was almost cut asunder, dragged overboard, and never seen afterward. The line was cut at that moment, but without avail. The fish descended to a considerable depth and died, from whence it was drawn to the surface by the lines connected with it and secured."

Whaling has almost ceased to have a place in the long list of our national industries. Its implements and the relics of old-time cruises fill niches in museums as memorials of a practically extinct calling. Along the wharves of New Bedford and New London a few old brigs lie rotting, but so effective have been the ravages of time that scarcely any of the once great fleet survive even in this invalid condition. The whales have been driven far into the Arctic regions, whither a few whalers employing the modern and unsportsmanlike devices of steam and explosives, follow them for a scanty profit. But the glory of the whale fishery is gone, leaving hardly a record behind it. In its time it employed thousands of stout sailors; it furnished the navy with the material that made that branch of our armed service the pride and glory of the nation. It explored unknown seas and carried the flag to undiscovered lands. Was not an Austrian exploring expedition, interrupted as it was about to take possession of land in the Antarctic in the name of Austria by encountering an American whaler, trim and trig, lying placidly at anchor in a harbor where the Austrian thought no man had ever been? It built up towns in New England that half a century of lethargy has been unable to kill. And so if its brigs—and its men—now molder, if its records are scanty and its history unwritten, still Americans must ever regard the whale fishery as one of the chief factors in the building of the nation—one of the most admirable chapters in our national story.

CHAPTER V

THE PRIVATEERS—PART TAKEN BY MERCHANT SAILORS IN BUILDING UP THE PRIVATEERING SYSTEM—LAWLESS STATE OF THE HIGH SEAS—METHOD OF DISTRIBUTING PRIVATEERING PROFITS—PICTURESQUE FEATURES OF THE CALLING—THE GENTLEMEN SAILORS—EFFECT ON THE REVOLUTIONARY ARMY—PERILS OF PRIVATEERING—THE OLD JERSEY PRISON SHIP—EXTENT OF PRIVATEERING—EFFECT ON AMERICAN MARINE ARCHITECTURE—SOME FAMOUS PRIVATEERS—THE "CHASSEUR," THE "PRINCE DE NEUFCHÂTEL," THE "MAMMOTH"—THE SYSTEM OF CONVOYS AND THE "RUNNING SHIPS"—A TYPICAL PRIVATEERS' BATTLE—THE "GENERAL ARMSTRONG" AT FAYAL—SUMMARY OF THE WORK OF THE PRIVATEERS.

In the early days of a new community the citizen, be he never so peaceful, is compelled, perforce, to take on the ways and the trappings of the fighting man. The pioneer is half hunter, half scout. The farmer on the outposts of civilization must be more than half a soldier; the cowboy or ranchman on our southwest frontier goes about a walking arsenal, ready at all times to take the laws into his own hands, and scorning to call on sheriffs or other peace officers for protection against personal injury. And while the original purpose of this militant, even defiant, attitude is self-protection, those who are long compelled to maintain it conceive a contempt for the law, which they find inadequate to guard them, and not infrequently degenerate into bandits.

It is hardly too much to say that the nineteenth century was already well into its second quarter before there was a semblance of recognized law upon the high seas. Pirates and buccaneers, privateers, and the naval vessels of the times that were little more than pirates, made the lot of the merchant sailor of the seventeenth and eighteenth centuries a precarious one. Wars were constant, declared on the flimsiest pretexts and with scant notice; so that the sailor putting out from port in a time of universal peace could feel no certainty that the first foreign vessel he met might not capture him as spoil of some war of which he had no knowledge. Accordingly, sailors learned to defend themselves, and the ship's armory was as necessary and vastly better stocked than the ship's medicine case. To point a carronade became as needful an accomplishment as to box the compass; and he was no A.B. who did not know how to swing a cutlass.

Out of such conditions, and out of the wars which the Napoleonic plague forced upon the world, sprung the practise of privateering; and while it is the purpose of this book to tell the story of the American merchant sailor only, it could not be complete without some account, however brief, of the American privateersman. For, indeed, the two were one throughout a considerable period of our maritime history, the sailor turning privateersman or the privateersman sailor as political or trade conditions demanded. In our colonial times, and in the earlier days of the nation, to be a famous privateersman, or to have had a hand in fitting out a successful privateer, was no mean passport to fame and fortune. Some of the names most eminent in the history of our country appear in connection with the outfitting or command of privateers; and not a few of the oldest fortunes of New England had their origin in this form of legalized piracy. And, after all, it is the need of the times that fixes the morality of an act. To-day privateering is dead; not by any formal agreement, for the United States, at the Congress of Paris, refused to agree to its outlawry; but in our war with Spain no recourse was had to letters of marque by either combatant, and it seems unlikely that in any future war between civilized nations either party will court the contempt of the world by going back to the old custom of chartering banditti to steal the property of private citizens of the hostile nation if found at sea. Private property on shore has long been respected by the armies of Christendom, and why its presence in a ship rather than in a cart makes it a fit object of plunder baffles the understanding. Perhaps in time the kindred custom of awarding prize money to naval officers, which makes of them a species of privateers, and pays them for capturing a helpless merchant ship, while an army officer gets nothing for taking the most powerful fort, may likewise be set aside as a relic of medieval warfare.

In its earliest days, of course, privateering was the weapon of a nation weak at sea against one with a large navy. So when the colonies threw down the gage of battle to Great Britain, almost the first act of the Revolutionary government was to authorize private owners to fit out armed ships to prey on British commerce. Some of the shipowners of New England had enjoyed some experience of the profits of this peculiar industry in the Seven Years' War, when quite a number of colonial privateers harried the French on the seas, and accordingly the response was prompt. In enterprises of this character the system of profit-sharing, already noted in connection with whaling, obtained. The owners took a certain share of each prize, and the remainder was divided among the officers and crew in certain fixed proportions. How great were the profits accruing to a privateersman in a "run of luck" might be illustrated by two facts set forth by Maclay, whose "History of American Privateers" is the chief authority on the subject. He asserts that "it frequently happened that even the common sailors received as their share in one cruise, over and above their wages, one thousand dollars—a small fortune in those days for a mariner," and further that "one of the boys in the 'Ranger,' who less than a month before had left a farm, received as his share one ton of sugar, from thirty to forty gallons of fourth-proof Jamaica rum, some twenty pounds of cotton, and about the same quantity of ginger, logwood, and allspice, besides seven hundred dollars in money." To be sure, in order to enjoy gains like these, the men had to risk the perils of battle in addition to the common ones of the sea; but it is a curious fact, recognized in all branches of industry, that the mere peril of a calling does not deter men from following it, and when it promises high profit it is sure to be overcrowded. In civil life to-day the most dangerous callings are those which are, as a rule, the most ill paid.

Very speedily the privateersmen became the most prosperous and the most picturesque figures along the waterside of the Atlantic cities. While the dignified merchant or shipowner, with a third interest in the "Daredevil" or the "Flybynight," might still maintain the sober demeanor of a good citizen and a pillar of the church, despite his profits of fifty or an hundred per cent. on each cruise, the gallant sailors who came back to town with pockets full of easily-won money, and the recollection of long and dismal weeks at sea behind them, were spectacular in their rejoicings. Their money was poured out freely while it lasted; and their example stirred all the townsboys, from the best families down to the scourings of the docks, to enter the same gentlemanlike profession.

Queerly enough, in a time of universal democracy, a provision was made on many of the privateers for the young men of family who desired to follow the calling. They were called "gentlemen sailors," and, in consideration of their social standing and the fact that they were trained to arms, were granted special and unusual privileges, such as freedom from the drudgery of working the ship, better fare than the common sailors, and more comfortable quarters. Indeed, they were free of duty except when fighting was to be done, and at other times fulfilled the function of the marine guards on our modern men-of-war. This came to be a very popular calling for adventurous young men of some family influence.

It has been claimed by some writers that "the Revolution was won by the New England privateers"; and, indeed, there can be no doubt that their activity did contribute in no small degree to the outcome of that struggle. Britain was then, as now, essentially a commercial nation, and the outcry of her merchants when the ravages of American privateers drove marine insurance rates up to thirty-three per cent., and even for a time made companies refuse it altogether, was clamorous. But there was another side to the story. Privateering, like all irregular service, was demoralizing, not alone to the men engaged in it, but to the youth of the country as well. The stories of the easy life and the great profits of the privateersmen were circulated in every little town, while the revels of these sea soldiers in the water-front villages were described with picturesque embellishments throughout the land. As a result, it became hard to get young men of spirit into the patriot armies. Washington complained that when the fortunes of his army were at their lowest, when he could not get clothing for his soldiers, and the snow at

Valley Forge was stained with the blood of their unshod feet, any American shipping on a privateer was sure of a competence, while great fortunes were being made by the speculators who fitted them out. Nor was this all. Such was the attraction of the privateer's life that it drew to it seamen from every branch of the maritime calling. The fisheries and the West India trade, which had long been the chief mainstay of New England commerce, were ruined, and it seemed for a time as if the hardy race of American seamen were to degenerate into a mere body of buccaneers, operating under the protection of international law, but plunderers and spoilers nevertheless. Fortunately, the long peace which succeeded the War of 1812 gave opportunity for the naturally lawful and civilized instincts of the Americans to assert themselves, and this peril was averted.

It is, then, with no admiration for the calling, and yet with no underestimate of its value to the nation, that I recount some of the achievements of those who followed it. The periods when American privateering was important were those of the Revolution and the War of 1812. During the Civil War the loss incurred by privateers fell upon our own people, and it is curious to note how different a tone the writers on this subject adopt when discussing the ravages of the Confederate privateers and those which we let loose upon British commerce in the brave days of 1812.

A true type of the Revolutionary privateersmen was Captain Silas Talbot, of Massachusetts. He was one of the New England lads apprenticed to the sea at an early age, having been made a cabin-boy at twelve. He rose to command and acquired means in his profession, as we have seen was common among our early merchant sailors, and when the Revolution broke out was living comfortably in his own mansion in Providence. He enlisted in Washington's army, but left it to become a privateer; and from that service he stepped to the quarter-deck of a man-of-war. This was not an uncommon line of development for the early privateersmen; and, indeed, it was not unusual to find navy officers, temporarily without commands, taking a cruise or two as privateers, until Congress should provide more ships for the regular service—a system which did not tend to make a Congress, which was niggardly at best, hasten to provide public vessels for work which was being reasonably well done at private expense. As a result of this system, we find such famous naval names as Decatur, Porter, Hopkins, Preble, Barry, and Barney also figuring in the lists of privateersmen. Talbot's first notable exploit was clearing New York harbor of several British men-of-war by the use of fire-ships. Washington, with his army, was then encamped at Harlem Heights, and the British ships were in the Hudson River menacing his flank. Talbot, in a fire-ship, well loaded with combustibles, dropped down the river and made for the biggest of the enemy's fleet, the "Asia." Though quickly discovered and made the target of the enemy's battery, he held his vessel on her course until fairly alongside of and entangled with the "Asia," when the fuses were lighted and the volcanic craft burst into roaring flames from stem to stern. So rapid was the progress of the flames that Talbot and his companions could scarcely escape with their lives from the conflagration they had themselves started, and he lay for days, badly burned and unable to see, in a little log hut on the Jersey shore. The British ships were not destroyed; but, convinced that the neighborhood was unsafe for them, they dropped down the bay; so the end sought for was attained. In 1779 Talbot was given command of the sloop "Argo," of 100 tons; "a mere shallop, like a clumsy Albany sloop," says his biographer. Sixty men from the army, most of whom had served afloat, were given him for crew, and he set out to clear Long Island Sound of Tory privateers; for the loyalists in New York were quite as avid for spoils as the New England Revolutionists. On his second cruise he took seven prizes, including two of these privateers. One of these was a 300-ton ship, vastly superior to the "Argo" in armament and numbers, and the battle was a fierce one. Nearly every man on the quarter-deck of the "Argo" was killed or wounded; the speaking trumpet in Talbot's hand was pierced by two bullets, and a cannon-ball carried away the tail of his coat. The damages sustained in this battle were scarce repaired when another British privateer appeared, and Talbot again went into action and took her, though of scarce half her size. In all this little "Argo"—which, by the way, belonged to Nicholas Low, of New York, an ancestor of the eminent Seth Low—took twelve prizes. Her commander was finally captured and sent first to the infamous "Jersey" prison-ship, and afterward to the Old Mill Prison in England.

NEARLY EVERY MAN ON THE QUARTERDECK OF THE "ARGO" WAS KILLED OR WOUNDED.

The "Jersey" prison-ship was not an uncommon lot for the bold privateersman, who, when once consigned to it, found that the reward of a sea-rover was not always wealth and pleasure. A Massachusetts privateersman left on record a contemporary account of the sufferings of himself and his comrades in this pestilential hulk, which may well be condensed here to show some of the perils that the adventurers dared when they took to the sea.

THE PRISON SHIP "JERSEY."

After about one-third of the captives made with this writer had been seized and carried away to serve against their country on British war-ships, the rest were conveyed to the "Jersey," which had been originally a 74-gun ship, then cut down to a hulk and moored at the Wallabout, at that time a lonely and deserted place on the Long Island shore, now about the center of the Brooklyn river front. "I found myself," writes the captive, "in a loathsome prison among a collection of the most wretched and disgusting objects I ever beheld in human form. Here was a motley crew covered with rags and filth, visages pallid with disease, emaciated with hunger and anxiety, and retaining hardly a trace of their original appearance.... The first day we could obtain no food, and seldom on the second could prisoners secure it in season for cooking it. Each prisoner received one-third as much as was allotted to a tar in the British navy. Our bill of fare was as follows: On Sunday, one pound of biscuit, one pound of pork, and half a pint of peas; Monday, one pound of biscuit, one pint of oatmeal, and two ounces of butter; Tuesday, one pound of biscuit and two pounds of salt beef, etc., etc. If this food had been of good quality and properly cooked, as we had no labor to perform, it would have kept us comfortable; but all our food appeared to be damaged. As for the pork, we were cheated out of more than half of it, and when it was obtained one would have judged from its motley hues, exhibiting the consistency and appearance of variegated fancy soap, that it was the flesh of the porpoise or sea-hog, and had been an inhabitant of the ocean rather than the sty. The peas were about as digestible as grape-shot; and the butter—had it not been for its adhesive properties to retain together the particles of biscuit that had been so riddled by the worms as to lose all their attraction of cohesion, we should not have considered it a desirable addition to our viands. The flour and oatmeal were sour, and the suet might have been nosed the whole length of our ship. Many times since, when I have seen in the country a large kettle of potatoes and pumpkins steaming over the fire to satisfy the appetite of some farmer's swine, I have thought of our destitute and starved condition, and what a luxury we should have considered the contents of that kettle aboard the 'Jersey.'... About two hours before sunset orders were given the prisoners to carry all their things below; but we were permitted to remain above until we retired for the night into our unhealthy and crowded dungeons. At sunset our ears were saluted with the insulting and hateful sound from our keepers of 'Down, rebels, down,' and we were hurried below, the hatchways fastened over us, and we were left to pass the night amid the accumulated horrors of sighs and groans, of foul vapor, a nauseous and putrid atmosphere, in a stifled and almost suffocating heat.... When any of the prisoners had died during the night, their bodies were brought to the upper deck in the morning and placed upon the gratings. If the deceased had owned a blanket, any prisoner might sew it around the corpse; and then it was lowered, with a rope tied round the middle, down the side of the ship into a boat. Some of the prisoners were allowed to go on shore under a guard to perform the labor of interment. In a bank near the Wallabout, a hole was excavated in the sand, in which the body was put, then slightly covered. Many bodies would, in a few days after this mockery of a burial, be exposed nearly bare by the action of the elements."

Such was, indeed, the end of many of the most gallant of the Revolutionary privateersmen; but squalid and cruel as was the fate of these unfortunates, it had no effect in deterring others from seeking fortune in the same calling. In 1775-76 there were commissioned 136 vessels, with 1360 guns; in 1777, 73 vessels, with 730 guns; in 1778, 115 privateers, with a total of 1150 guns; in 1779, 167 vessels, with 2505 guns; in 1780, 228 vessels, with 3420 guns; in 1781, 449 vessels, with 6735 (the high-water mark): and in 1782, 323 vessels, with 4845 guns. Moreover, the vessels grew in size and efficiency, until toward the latter end of the war they were in fact well-equipped war-vessels, ready to give a good account of themselves in a fight with a British frigate, or even to engage a shore battery and cut out prizes from a hostile harbor. It is, in fact, a striking evidence of the gallantry and the patriotism of the privateersmen that they did not seek to evade battle with the enemy's armed forces. Their business was, of course, to earn profits for the merchants who had fitted them out, and profits were most easily earned by preying upon inferior or defenseless vessels. But the spirit of the war was strong upon many of them, and it is not too much to say that the privateers were handled as gallantly and accepted unfavorable odds in battle as readily as could any men-of-war. Their ravages upon British commerce plunged all commercial England into woe. The war had hardly proceeded two years when it was formally declared in the House of Commons that the losses to American privateers amounted to seven hundred and thirty-three ships, of a value of over $11,000,000. Mr. Maclay estimates from this that "our amateur man-of-war's men averaged more than four prizes each," while some took twenty and one ship twenty-eight in a single cruise. Nearly eleven hundred prisoners were taken with the captured ships. While there are no complete figures for the whole period of the war obtainable, it is not to be believed that quite so high a record was maintained, for dread of privateers soon drove British shipping into their harbors, whence they put forth, if at all, under the protection of naval convoys. Nevertheless, the number of captures must have continued great for some years; for, as is shown by the foregoing figures, the spoils were sufficiently attractive to cause a steady increase in the number of privateers until the last year of the war.

There followed dull times for the privateersmen. Most of them returned to their ordinary avocations of sea or shore—became peaceful sailors, or fishermen, or ship-builders, or farmers once again. But in so great a body of men who had lived sword in hand for years, and had fattened on the spoils of the commerce of a great nation, it was inevitable that there should be many utterly unable to return to the humdrum life of honest industry. Many drifted down to that region of romance and outlawry, dear to the heart of the romantic boy, the Spanish Main, and there, as pirates in a small way and as buccaneers, pursued the predatory life. For a time the war which sprung up between England and France seemed to promise these turbulent spirits congenial and lawful occupation. France, it will be remembered, sent the Citizen Genet over to the United States to take advantage of the supposed gratitude of the American people for aid during the Revolution to fit out privateers and to make our ports bases of operation against the British. It must be admitted that Genet would have had an easy task, had he had but the people to reckon with. He found privateering veterans by the thousand eager to take up that manner of life once more. In all the seacoast towns were merchants quite as ready for profitable ventures in privateering under the French flag as under their own, provided they could be assured of immunity from governmental prosecution. And, finally, he found the masses of the people fired with enthusiasm for the principles of the French Revolution, and eager to show sympathy for a people who, like themselves, had thrown off the yoke of kings. The few privateers that Minister Genet fitted out before President Washington became aroused to his infraction of the principles of neutrality were quickly manned, and began sending in prizes almost before they were out of sight of the American shore. The crisis came, however, when one of these ships actually captured a British merchantman in Delaware Bay. Then the administration made a vigorous protest, demanded the release of the vessels taken, arrested two American sailors who had shipped on the privateer, and broke up at once the whole project of the Frenchman. It was a critical moment in our national history, for, between France and England abroad, the Federalist and Republican at home, the President had to steer a course beset with reefs. The maritime community was not greatly in sympathy with his suppression of the French minister's plans, and with some reason, for British privateers had been molesting our vessels all along our coasts and distant waters. It was a time when no merchant could tell whether the stout ship he had sent out was even then discharging her cargo at her destination, or tied up as a prize in some British port. We Americans are apt to regard with some pride Washington's stout adherence to the most rigid letter of the law of neutrality in those troublous times, and our historians have been at some pains to impress us with the impropriety of Jefferson's scarcely concealed liking for France; but the fact is that no violation of the neutrality law which Genet sought was more glaring than those continually committed by Great Britain, and which our Government failed to resent. In time France, moved partly by pique because of our refusal to aid her, and partly by contempt for a nation that failed to protect its ships against British aggression, began itself to prey upon our commerce. Then the state of our maritime trade was a dismal one. Our ships were the prey of both France and England; but since we were neutral, the right of fitting out privateers of our own was denied our shipping interests. We were ground between the upper and nether millstones.

But, as so often happens, persecution bred the spirit and created the weapons for its correction. When it was found that every American vessel was the possible spoil of any French or English cruiser or privateer that she might encounter; that our Government was impotent to protect its seamen; that neither our neutrality rights nor the neutrality of ports in which our vessels lay commanded the respect of the two great belligerents, the Yankee shipping merchants set about meeting the situation as best they might. They did not give up their effort to secure the world's trade—that was never an American method of procedure. But they built their ships so as to be able to run away from anything they might meet; and they manned and armed them so as to fight if fighting became necessary. So the American merchantman became a long, sharp, clipper-built craft that could show her heels to almost anything afloat; moderate of draft, so that she could run into lagoons and bays where no warship could follow. They mounted from four to twelve guns, and carried an armory of rifles and cutlasses which their men were well trained to handle. Accordingly, when the depredations of foreign nations became such as could not longer be borne, and after President Jefferson's plan of punishing Europe for interfering with our commerce by laying an embargo which kept our ships at home had failed, war was declared with England; and from every port on the Atlantic seaboard privateers—ships as fit for their purpose as though specially built for it—swarmed forth seeking revenge and spoils. Their very names told of the reasons of the American merchantmen for complaint—the reasons why they rejoiced that they were now to have their turn. There were the "Orders-in-Council," the "Right-of-Search," the "Fair-trader," the "Revenge." Some were mere pilot-boats, with a Long Tom amidships and a crew of sixty men; others were vessels of 300 tons, with an armament and crew like a man-of-war. Before the middle of July, 1812, sixty-five such privateers had sailed, and the British merchantmen were scudding for cover like a covey of frightened quail.

The War of 1812 was won, so far as it was won at all, on the ocean. In the land operations from the very beginning the Americans came off second best; and the one battle of importance in which they were the victors—the battle of New Orleans—was without influence upon the result, having been fought after the treaty of peace had been signed at Ghent. But on the ocean the honors were all taken by the Americans, and no small share of these honors fell to the private armed navy of privateers. As the war progressed these vessels became in type more like the regular sloop-of-war, for the earlier craft, while useful before the British began sending out their merchantmen under convoy, proved to be too small to fight and too light to escape destruction from one well-aimed broadside. The privateer of 1813 was usually about 115 to 120 feet long on the spar-deck, 31 feet beam, and rigged as a brig or ship. They were always fast sailers, and notable for sailing close to the wind. While armed to fight, if need be, that was not their purpose, and a privateersman who gained the reputation among owners of being a fighting captain was likely to go long without a command. Accordingly, these vessels were lightly built and over-rigged (according to the ideas of British naval construction), for speed was the great desideratum. They were at once the admiration and the envy of the British, who imitated their models without success and tried to utilize them for cruisers when captured, but destroyed their sailing qualities by altering their rig and strengthening their hulls at the expense of lightness and symmetry.

I have already referred to Michael Scott's famous story of sea life, "Tom Cringle's Log," which, though in form a work of fiction, contains so many accounts of actual happenings, and expresses so fully the ideas of the British naval officer of that time, that it may well be quoted in a work of historical character. Tom Cringle, after detailing with a lively description the capture of a Yankee privateer, says that she was assigned to him for his next command. He had seen her under weigh, had admired her trim model, her tapering spars, her taut cordage, and the swiftness with which she came about and reached to windward. He thus describes the change the British outfitters made in her:

"When I had last seen her she was the most beautiful little craft, both in hull and rigging, that ever delighted the eyes of a sailor; but the dock yard riggers and carpenters had fairly bedeviled her at least so far as appearances went. First, they had replaced the light rail on her gunwale by heavy, solid bulwarks four feet high, surmounted by hammock nettings at least another foot; so that the symmetrical little vessel, that formerly floated on the foam light as a seagull, now looked like a clumsy, dish-shaped Dutch dogger. Her long, slender wands of masts, which used to swing about as if there were neither shrouds nor stays to support them, were now as taut and stiff as church-steeples, with four heavy shrouds of a side, and stays, and back-stays, and the devil knows what all."

It is a curious fact that no nation ever succeeded in imitating these craft. The French went into privateering without in the least disturbing the equanimity of the British shipowner; but the day the Yankee privateers took the sea a cry went up from the docks and warehouses of Liverpool and London that reverberated among the arches of Westminster Hall. The newspapers were loud in their attacks upon the admiralty authorities. Said the *Morning Chronicle* in 1814:

"That the whole coast of Ireland, from Wexford round by Cape Clear to Carrickfergus, should have been for above a month under the unresisted domination of a few petty fly-by-nights from the blockaded ports of the United States is a grievance equally intolerable and disgraceful."

This wail may have resulted from the pleasantry of one Captain Boyle, of the privateer "Chasseur," a famous Baltimore clipper, mounting sixteen guns, with a complement of one hundred officers, seamen, and marines. Captain Boyle, after exhausting, as it seemed to him, the possibilities of the West Indies for excitement and profit, took up the English channel for his favorite cruising-ground. One of the British devices of that day for the embarrassment of an enemy was what is called a "paper blockade." That is to say, when it appeared that the blockading fleet had too few vessels to make the blockade really effective by watching each port, the admiral commanding would issue a proclamation that such and such ports were in a state of blockade, and then withdraw his vessels from those ports; but still claim the right to capture any neutral vessels which he might encounter bound thither. This practise is now universally interdicted by international law, which declares that a blockade, to be binding upon neutrals, must be effective. But in those days England made her own international law—for the sea, at any rate—and the paper blockade was one of her pet weapons. Captain Boyle satirized this practise by drawing up a formal proclamation of blockade of all the ports of Great Britain and Ireland, and sending it to Lloyds, where it was actually posted. His action was not wholly a jest, either, for he did blockade the port of St. Vincent so effectively for five days that the inhabitants sent off a pitiful appeal to Admiral Durham to send a frigate to their relief.

It was at this time, too, that the *Annual Register* recorded as "a most mortifying reflection" that, with a navy of more than one thousand ships in commission, "it was not safe for a British vessel to sail without convoy from one part of the English or Irish Channel to another." Merchants held meetings, insurance corporations and boards of trade memorialized the government on the subject; the shipowners and merchants of Glasgow, in formal resolutions, called the attention of the admiralty to the fact that "in the short space of twenty-four months above eight hundred vessels have been captured by the power whose maritime strength we have hitherto impolitically held in contempt." It was, indeed, a real blockade of the British Isles that was effected by these irregular and pigmy vessels manned by the sailors of a nation that the British had long held in high scorn. The historian Henry Adams, without attempting to give any complete list of captures made on the British coasts in 1814, cites these facts:

"The 'Siren,' a schooner of less than 200 tons, with seven guns and seventy-five men, had an engagement with His Majesty's cutter 'Landrail,' of four guns, as the cutter was crossing the Irish sea with dispatches. The 'Landrail' was captured, after a somewhat smart action, and was sent to America, but was recaptured on the way. The victory was not remarkable, but the place of capture was very significant, and it happened July 12 only a fortnight after Blakely captured the 'Reindeer' farther westward. The 'Siren' was but one of many privateers in those waters. The 'Governor Tompkins' burned fourteen vessels successively in the British Channel. The 'Young Wasp,' of Philadelphia, cruised nearly six months about the coasts of England and Spain, and in the course of West India commerce. The 'Harpy,' of Baltimore, another large vessel of some 350 tons and fourteen guns, cruised nearly three months off the coast of Ireland, in the British Channel, and in the Bay of Biscay, and returned safely to Boston filled with plunder, including, as was said, upward of £100,000 in British treasury notes and bills of exchange. The 'Leo,' a Boston schooner of about 200 tons, was famous for its exploits in these waters, but was captured at last by the frigate 'Tiber,' after a chase of about eleven hours. The 'Mammoth,' a Baltimore schooner of nearly 400 tons, was seventeen days off Cape Clear, the southernmost point of Ireland. The most mischievous of all was the 'Prince of Neufchâtel,' New York, which chose the Irish Channel as its favorite haunt, where during the summer it made ordinary coasting traffic impossible."

The vessels enumerated by Mr. Adams were by no means among the more famous of the privateers of the War of 1812; yet when we come to examine their records we find something notable or something romantic in the career of each—a fact full of suggestion of the excitement of the privateersman's life. The "Leo," for example, at this time was under command of Captain George Coggeshall, the foremost of all the privateers, and a man who so loved his calling that he wrote an excellent book about it. Under an earlier commander she made several most profitable cruises, and when purchased by Coggeshall's associates was lying in a French port. France and England were then at peace, and it may be that the French remembered the way in which we had suppressed the Citizen Genet. At any rate, they refused to let Coggeshall take his ship out of the harbor with more than one gun—a Long Tom—aboard. Nothing daunted, he started out with this armament, to which some twenty muskets were added, on a privateering cruise in the channel, which was full of British cruisers. Even the Long Tom proved untrustworthy, so recourse was finally had to carrying the enemy by boarding; and in this way four valuable prizes were taken, of which three were sent home with prize crews. But a gale carried away the "Leo's" foremast, and she fell a prey to an English frigate which happened along untimely.

The "Mammoth" was emphatically a lucky ship. In seven weeks she took seventeen merchantmen, paying for herself several times over. Once she fought a lively battle with a British transport carrying four hundred men, but prudently drew off. True, the Government was paying a bonus of twenty-five dollars a head for prisoners; but cargoes were more valuable. Few of the privateers troubled to send in their prisoners, if they could parole and release them. In all, the "Mammoth" captured twenty-one vessels, and released on parole three hundred prisoners.

Of all the foregoing vessels, the "Prince de Neufchâtel" was the most famous. She was an hermaphrodite brig of 310 tons, mounting 17 guns. She was a "lucky" vessel, several times escaping a vastly superior force and bringing into port, for the profit of her owners, goods valued at $3,000,000, besides large quantities of specie. Her historic achievement, however, was beating off the British frigate "Endymion," off Nantucket, one dark night, after a battle concerning which a British naval historian, none too friendly to Americans, wrote: "So determined and effective a resistance did great credit to the American captain and his crew." The privateer had a prize in tow, by which, of course, her movements were much hampered, for her captain was not inclined to save himself at the expense of his booty. But, more than this, she had thirty-seven prisoners aboard, while her own crew was sorely reduced by manning prizes. The night being calm, the British attempted to take the ship by boarding from small boats, for what reason does not readily appear, since the vessels were within range of each other, and the frigate's superior metal could probably have reduced the Americans to subjection. Instead, however, of opening fire with his broadside, the enemy sent out boarding parties in five boats. Their approach was detected on the American vessel, and a rapid fire with small arms and cannon opened upon them, to which they paid no attention, but pressed doggedly on. In a moment the boats surrounded the privateer—one on each bow, one on each side, and one under the stern—and the boarders began to swarm up the sides like cats. It was a bloody hand-to-hand contest that followed, in which every weapon, from cutlass and clubbed musket down to bare hands, was employed. Heavy shot, which had been piled up in readiness on deck, were thrown into the boats in an effort to sink them. Hundreds of loaded muskets were ranged along the rail, so that the firing was not interrupted to reload. Time and again the British renewed their efforts to board, but were hurled back by the American defenders. A few who succeeded in reaching the decks were cut down before they had time to profit by their brief advantage. Once only did it seem that the ship was in danger. Then the assailants, who outnumbered the Americans four to one, had reached the deck over the bows in such numbers that they were gradually driving the defenders aft. Every moment more men came swarming over the side; and as the Americans ran from all parts of the ship to meet and overpower those who had already reached the deck, new ways were opened for others to clamber aboard. The situation was critical; but was saved by Captain Ordronaux by a desperate expedient, and one which it is clear would have availed nothing had not his men known him for a man of fierce determination, ready to fulfil any desperate threat. Seizing a lighted match from one of the gunners, he ran to the hatch immediately over the magazine, and called out to his men that if they retreated farther he would blow up the ship, its defenders, and its assailants. The men rallied. They swung a cannon in board so that it commanded the deck, and swept away the invaders with a storm of grape. In a few minutes the remaining British were driven back to their boats. The battle had lasted less than half an hour when the British called for quarter, the smoke cleared away, the cries of combat ceased, and both parties were able to count their losses. The crew of the privateer had numbered thirty-seven, of whom seven were killed and twenty-four wounded. The British had advanced to the attack with a force of one hundred and twenty-eight, in five boats. Three of the boats drifted away empty, one was sunk, and one was captured. Of the attacking force not one escaped; thirty were made prisoners, many of them sorely wounded, and the rest were either killed or swept away by the tide and drowned. The privateers actually had more prisoners than they had men of their own. Some of the prisoners were kept towing in a launch at the stern, and, by way of strategy, Captain Ordronaux set two boys to playing a fife and drum and stamping about in a sequestered part of his decks as though he had a heavy force aboard. Only by sending the prisoners ashore under parole was the danger of an uprising among the captives averted.

IF THEY RETREATED FARTHER HE WOULD BLOW UP THE SHIP

In the end the "Prince de Neufchâtel" was captured by a British squadron, but only after a sudden squall had carried away several of her spars and made her helpless.

As the war progressed it became the custom of British merchants to send out their ships only in fleets, convoyed by one or two men-of-war, a system that, of course, could be adopted only by nations very rich in war-ships. The privateers' method of meeting this was to cruise in couples, a pair of swift, light schooners, hunting the prize together. When the convoy was encountered, both would attack, picking out each its prey. The convoys were usually made up with a man-of-war at the head of the column, and as this vessel would make sail after one of the privateers, the other would rush in at some point out of range, and cut out its prize. When the British began sending out two ships of war with each convoy, the privateers cruised in threes, and the same tactics were observed.

But the richest prizes won by the privateer were the single going ships, called "running ships," that were prepared to defend themselves, and scorned to wait for convoy. These were generally great packets trading to the Indies, whose cargoes were too valuable to be delayed until some man-of-war could be found for their protection. They were heavily armed, often, indeed, equaling a frigate in their batteries and the size of their crews. But, although to attack one of these meant a desperate fight, the Yankee privateer always welcomed the chance, for besides a valuable cargo, they were apt to carry a considerable sum in specie. The capture of one of these vessels, too, was the cause of annoyance to the enemy disproportionate to even their great value to their captors, for they not only carried the Royal Mail, but were usually the agencies by which the dispatches of the British general were forwarded. Mail and dispatches, alike, were promptly thrown overboard by their captors.

In the diary of a privateersman of Revolutionary days is to be found the story of the capture of an Indiaman which may well be reprinted as typical.

"I THINK SHE IS A HEAVY SHIP."

"As the fog cleared up, we perceived her to be a large ship under English colors, to the windward, standing athwart our starboard bow. As she came down upon us, she appeared as large as a seventy-four; and we were not deceived respecting her size, for it afterwards proved that she was an old East Indiaman, of 1100 tons burden, fitted out as a letter of marque for the West India trade, mounted with thirty-two guns, and furnished with a complement of one hundred and fifty men. She was called the 'Admiral Duff,' commanded by Richard Strange, from St. Christopher and St. Eustachia, laden with sugar and tobacco, and bound to London. I was standing near our first lieutenant, Mr. Little, who was calmly examining the enemy as she approached, with his spy-glass, when Captain Williams stepped up and asked his opinion of her. The lieutenant applied the glass to his eye again and took a deliberate look in silence, and replied: 'I think she is a heavy ship, and that we shall have some hard fighting, but of one thing I am certain, she is not a frigate; if she were, she would not keep yawing and showing her broadsides as she does; she would show nothing but her head and stern; we shall have the advantage of her, and the quicker we get alongside the better.' Our captain ordered English colors to be hoisted, and the ship to be cleared for action.

"The enemy approached 'till within musket-shot of us. The two ships were so near to each other that we could distinguish the officers from the men; and I particularly noticed the captain on the gangway, a noble-looking man, having a large gold-laced cocked hat on his head, and a speaking-trumpet in his hand. Lieutenant Little possessed a powerful voice, and he was directed to hail the enemy; at the same time the quartermaster was ordered to stand ready to haul down the English flag and to hoist up the American. Our lieutenant took his station on the after part of the starboard gangway, and elevating his trumpet, exclaimed: 'Hullo. Whence come you?'

"'From Jamaica, bound to London,' was the answer.

"'What is the ship's name?' inquired the lieutenant.

"'The "Admiral Duff",' was the reply.

"The English captain then thought it his turn to interrogate, and asked the name of our ship. Lieutenant Little, in order to gain time, put the trumpet to his ear, pretending not to hear the question. During the short interval thus gained, Captain Williams called upon the gunner to ascertain how many guns could be brought to bear upon the enemy. 'Five,' was the answer. 'Then fire, and shift the colors,' were the orders. The cannons poured forth their deadly contents, and, with the first flash, the American flag took the place of the British ensign at our masthead.

"The compliment was returned in the form of a full broadside, and the action commenced. I was stationed on the edge of the quarter-deck, to sponge and load a six-pounder; this position gave me a fine opportunity to see the whole action. Broadsides were exchanged with great rapidity for nearly an hour; our fire, as we afterward ascertained, produced a terrible slaughter among the enemy, while our loss was as yet trifling. I happened to be looking for a moment toward the main deck, when a large shot came through our ship's side and killed a midshipman. At this moment a shot from one of our marines killed the man at the wheel of the enemy's ship, and, his place not being immediately supplied, she was brought alongside of us in such a manner as to bring her bowsprit directly across our forecastle. Not knowing the cause of this movement, we supposed it to be the intention of the enemy to board us. Our boarders were ordered to be ready with their pikes to resist any such attempt, while our guns on the main deck were sending death and destruction among the crew of the enemy. Their principal object now seemed to be to get liberated from us, and by cutting away some of their rigging, they were soon clear, and at the distance of a pistol shot.

"The action was then renewed, with additional fury; broadside for broadside continued with unabated vigor; at times, so near to each other that the muzzles of our guns came almost in contact, then again at such a distance as to allow of taking deliberate aim. The contest was obstinately continued by the enemy, although we could perceive that great havoc was made among them, and that it was with much difficulty that their men were compelled to remain at their quarters. A charge of grape-shot came in at one of our portholes, which dangerously wounded four or five of our men, among whom was our third lieutenant, Mr. Little, brother to the first.

"The action had now lasted about an hour and a half, and the fire from the enemy began to slacken, when we suddenly discovered that all the sails on her mainmast were enveloped in a blaze. Fire spread with amazing rapidity, and, running down the after rigging, it soon communicated with her magazine, when her whole stern was blown off, and her valuable cargo emptied into the sea. Our enemy's ship was now a complete wreck, though she still floated, and the survivors were endeavoring to save themselves in the only boat that had escaped the general destruction. The humanity of our captain urged him to make all possible exertions to save the miserable wounded and burned wretches, who were struggling for their lives in the water. The ship of the enemy was greatly our superior in size, and lay much higher out of the water. Our boats had been exposed to his fire, as they were placed on spars between the fore and mainmasts during the action, and had suffered considerable damage. The carpenters were ordered to repair them with the utmost expedition, and we got them out in season to take up fifty-five men, the greater part of whom had been wounded by our shot, or burned when the powder-magazine exploded. Their limbs were mutilated by all manner of wounds, while some were burned to such a degree that the skin was nearly flayed from their bodies. Our surgeon and his assistants had just completed the task of dressing the wounds of our own crew, and then they directed their attention to the wounded of the enemy. Several of them suffered the amputation of their limbs, five of them died of their wounds, and were committed to their watery graves. From the survivors we learned that the British commander had frequently expressed a desire to come in contact with a 'Yankee frigate' during his voyage, that he might have a prize to carry to London. Poor fellow. He little thought of losing his ship and his life in an engagement with a ship so much inferior to his own—with an enemy upon whom he looked with so much contempt."

But most notable of all the battles fought by privateersmen in the War of 1812, was the defense of the brig "General Armstrong," in the harbor of Fayal, in September, 1814. This famous combat has passed into history, not only because of the gallant fight made by the privateer, but because the three British men-of-war to whom she gave battle, were on their way to cooperate with Packenham at New Orleans, and the delay due to the injuries they received, made them too late to aid in that expedition, and may have thus contributed to General Jackson's success.

The "General Armstrong" had always been a lucky craft, and her exploits in the capture of merchantmen, no less than the daring of her commander in giving battle to ships-of-war which he encountered, had won her the peculiar hate of the British navy. At the very beginning of her career, when in command of Captain Guy R. Champlin, she fought a British frigate for more than an hour, and inflicted such grave damage that the enemy was happy enough to let her slip away when the wind freshened. On another occasion she engaged a British armed ship of vastly superior strength, off the Surinam River, and forced her to run ashore. Probably the most valuable prize taken in the war fell to her guns—the ship "Queen," with a cargo invoiced at £90,000. Indeed, such had been her audacity, and so many her successes, that the British were eager for her capture or destruction, above that of any other privateer.

In September, 1814, the "General Armstrong," now under command of Captain Samuel G. Reid, was at anchor in the harbor at Fayal, a port of Portugal, when her commander saw a British war-brig come nosing her way into the harbor. Soon after another vessel appeared, and then a third, larger than the first two, and all flying the British ensign. Captain Reid immediately began to fear for his safety. It was true that he was in a neutral port, and under the law of nations exempt from attack, but the British had never manifested that extreme respect for neutrality that they exacted of President Washington when France tried to fit out privateers in our ports. More than once they had attacked and destroyed our vessels in neutral ports, and, indeed, it seemed that the British test of neutrality was whether the nation whose flag was thus affronted, was able or likely to resent it. Portugal was not such a nation.

All this was clear to Captain Reid, and when he saw a rapid signaling begun between the three vessels of the enemy, he felt confident that he was to be attacked. He had already discovered that the strangers were the 74-gun ship of the line "Plantagenet," the 38-gun frigate "Rota," and the 18-gun war-brig "Carnation," comprising a force against which he could not hope to win a victory. The night came on clear, with a bright moon, and as the American captain saw boats from the two smaller vessels rallying about the larger one, he got out his sweeps and began moving his vessel inshore, so as to get under the guns of the decrepit fort, with which Portugal guarded her harbor. At this, four boats crowded with men, put out from the side of the British ship, and made for the privateer, seeing which, Reid dropped anchor and put springs on his cables, so as to keep his broadside to bear on the enemy as they approached. Then he shouted to the British, warning them to keep off, or he would fire. They paid no attention to the warning, but pressed on, when he opened a brisk fire upon them. For a time there was a lively interchange of shots, but the superior marksmanship of the Americans soon drove the enemy out of range with heavy casualties. The British retreated to their ships with a hatred for the Yankee privateer even more bitter than that which had impelled them to the lawless attack, and a fiercer determination for her destruction.

It is proper to note, that after the battle was fought, and the British commander had calmly considered the possible consequences of his violation of the neutrality laws, he attempted to make it appear that the Americans themselves were the aggressors. His plea, as made in a formal report to the admiralty, was that he had sent four boats to discover the character of the American vessel; that they, upon hailing her, had been fired upon and suffered severe loss, and that accordingly he felt that the affront to the British flag could only be expiated by the destruction of the vessel. The explanation was not even plausible, for the British commander, elsewhere in his report, acknowledged that he was perfectly informed as to the identity of the vessel, and even had this not been the case, it is not customary to send four boats heavily laden with armed men, merely to discover the character of a ship in a friendly port.

The withdrawal of the British boats gave Captain Reid time to complete the removal of his vessel to a point underneath the guns of the Portuguese battery. This gave him a position better fitted for defense, although his hope that the Portuguese would defend the neutrality of their port, was destined to disappointment, for not a shot was fired from the battery.

"STRIVING TO REACH HER DECKS AT EVERY POINT"

Toward midnight the attack was resumed, and by this time the firing within the harbor had awakened the people of the town, who crowded down to the shore to see the battle. The British, in explanation of the reverse which they suffered, declared that all the Americans in Fayal armed themselves, and from the shore supplemented the fire from the "General Armstrong." Captain Reid, however, makes no reference to this assistance. In all, some four hundred men joined in the second attack. Twelve boats were in line, most of them with a howitzer mounted in the bow. The Americans used their artillery on these craft as they approached, and inflicted great damage before the enemy were in a position to board. The British vessels, though within easy gun-fire, dared not use their heavy cannon, lest they should injure their own men, and furthermore, for fear that the shot would fall into the town. The midnight struggle was a desperate one, the enemy fairly surrounding the "General Armstrong," and striving to reach her decks at every point. But though greatly outnumbered, the defenders were able to maintain their position, and not a boarder succeeded in reaching the decks. The struggle continued for nearly three-quarters of an hour, after which the British again drew off. Two boats filled with dead and dying men, were captured by the Americans, the unhurt survivors leaping overboard and swimming ashore. The British report showed, that in these two attacks there were about one hundred and forty of the enemy killed, and one hundred and thirty wounded. The Americans had lost only two killed and seven wounded, but the ship was left in no condition for future defense. Many of the guns were dismounted, and the Long Tom, which had been the mainstay of the defense, was capsized. Captain Reid and his officers worked with the utmost energy through the night, trying to fit the vessel for a renewal of the combat in the morning, but at three o'clock he was called ashore by a note from the American consul. Here he was informed that the Portuguese Governor had made a personal appeal to the British commander for a cessation of the attack, but that it had been refused, with the statement that the vessel would be destroyed by cannon-fire from the British ships in the morning. Against an attack of this sort it was, of course, futile for the "General Armstrong" to attempt to offer defense, and accordingly Captain Reid landed his men with their personal effects, and soon after the British began fire in the morning, scuttled the ship and abandoned her. He led his men into the interior, seized on an abandoned convent, and fortifying it, prepared to resist capture. No attempt, however, was made to pursue him, the British commander contenting himself with the destruction of the privateer. For nearly a week the British ships were delayed in the harbor, burying their dead and making repairs. When they reached New Orleans, the army which they had been sent to reenforce, had met Jackson on the plains of Chalmette, and had been defeated. The price paid for the "General Armstrong" was, perhaps, the heaviest of the war. The British commander seemed to appreciate this fact, for every effort was made to keep the news of the battle from becoming known in England, and when complete concealment was no longer possible, an official report was given out that minimized the British loss, magnified the number of the Americans, and totally mis-stated the facts bearing on the violation of the neutrality of the Portuguese port. Captain Reid, however, was made a hero by his countrymen. A Portuguese ship took him and his crew to Amelia Island, whence they made their way to New York. Poughkeepsie voted him a sword. Richmond citizens gave him a complimentary dinner, at which were drunk such toasts as: "The private cruisers of the United States—whose intrepidity has pierced the enemy's channels and bearded the lion in his den"; "Neutral Ports—whenever the tyrants of the ocean dare to invade these sanctuaries, may they meet with an 'Essex' and an 'Armstrong'"; and "Captain Reid—his valor has shed a blaze of renown upon the character of our seamen, and won for himself a laurel of eternal bloom." The newspapers of the times rang with eulogies of Reid, and anecdotes of his seafaring experiences. But after all, as McMaster finely says in his history: "The finest compliment of all was the effort made in England to keep the details of the battle from the public, and the false report of the British commander."

In finally estimating the effect upon the American fortunes in the War of 1812, of the privateers and their work, many factors must be taken into consideration. At first sight it would seem that a system which gave the services of five hundred ships and their crews to the task of annoying the British, and inflicting damage upon their commerce without cost to the American Government, must be wholly advantageous. We have already seen the losses inflicted upon British commerce by our privateers reflected in the rapidly increasing cost of marine insurance. While the statistics in the possession of the Government are not complete, they show that twenty-five hundred vessels at least were captured during the War of 1812 by these privately-owned cruisers, and there can be no shadow of a doubt that the loss inflicted upon British merchants, and the constant state of apprehension for the safety of their vessels in which they were kept, very materially aided in extending among them a willingness to see peace made on almost any terms.

But this is the other side of the story: The prime purpose of the privateer was to make money for its owners, its officers, and its crew. The whole design and spirit of the calling was mercenary. It inflicted damage on the enemy, but only incidentally to earning dividends for its participants. If Government cruisers had captured twenty-five hundred British vessels, those vessels would have been lost to the enemy forever. But the privateer, seeking gains, tried to send them into port, however dangerous such a voyage might be, and accordingly, rather more than a third of them were recaptured by the enemy. We may note here in passing, that one reason why the so-called Confederate privateers during our own Civil War, did an amount of damage so disproportionate to their numbers, was that they were not, in fact, privateers at all. They were commissioned by the Confederate Government to inflict the greatest possible amount of injury upon northern commerce, and accordingly, when Semmes or Maffitt captured a United States vessel, he burned it on the spot. There was no question of profit involved in the service of the "Alabama," the "Florida," or the "Shenandoah," and they have been called privateers in our histories, mainly because Northern writers have been loath to concede, to what they called a rebel government, the right to equip and commission regular men-of-war.

But to return to the American privateers of 1812. While, as I have pointed out, there were many instances of enormous gains being made, it is probable that the business as a whole, like all gambling businesses as a whole, was not profitable. Some ships made lucky voyages, but there is on record in the Navy Department a list of three hundred vessels that took not one single prize in the whole year of 1813. The records of Congress show that, as a whole, the business was not remunerative, because there were constant appeals from people interested. In response to this importunity, Congress at one time paid a bounty of twenty-five dollars a head for all prisoners taken. At other times it reduced the import duties on cargoes captured and landed by privateers. Indeed, it is estimated by a careful student, that the losses to the Government in the way of direct expenditures and remission of revenues through the privateering system, amounted to a sum sufficient to have kept twenty sloops of war on the sea throughout the period of hostilities, and there is little doubt that such vessels could have actually accomplished more in the direction of harassing the enemy than the privateers. A very grave objection to the privateering system, however, was the fact that the promise of profit to sailors engaged in it was so great, that all adventurous men flocked into the service, so that it became almost impossible to maintain our army or to man our ships. I have already quoted George Washington's objections to the practise during the Revolution. During the War of 1812, some of our best frigates were compelled to sail half manned, while it is even declared that the loss of the "Chesapeake" to the "Shannon" was largely due to the fact that her crew were discontented and preparing, as their time of service was nearly up, to quit the Government service for privateering. In a history of Marblehead, one of the famous old seafaring towns of Massachusetts, it is declared that of nine hundred men of that town who took part in the war, fifty-seven served in the army, one hundred and twenty entered the navy, while seven hundred and twenty-six shipped on the privateers. These figures afford a fair indication of the way in which the regular branches of the service suffered by the competition of the system of legalized piracy.

CHAPTER VI.

THE ARCTIC TRAGEDY—AMERICAN SAILORS IN THE FROZEN DEEP—THE SEARCH FOR SIR JOHN FRANKLIN—REASONS FOR SEEKING THE NORTH POLE—TESTIMONY OF SCIENTISTS AND EXPLORERS—PERTINACITY OF POLAR VOYAGERS—DR. KANE AND DR. HAYES—CHARLES F. HALL, JOURNALIST AND EXPLORER—MIRACULOUS ESCAPE OF HIS PARTY—THE ILL-FATED "JEANNETTE" EXPEDITION—SUFFERING AND DEATH OF DE LONG AND HIS COMPANIONS—A PITIFUL DIARY—THE GREELY EXPEDITION—ITS CAREFUL PLAN AND COMPLETE DISASTER—RESCUE OF THE GREELY SURVIVORS—PEARY, WELLMAN, AND BALDWIN.

A chapter in the story of the American sailor, which, though begun full an hundred years ago, is not yet complete, is that which tells the narrative of the search for the North Pole. It is a story of calm daring, of indomitable pertinacity, of patient endurance of the most cruel suffering, of heroic invitation to and acceptance of death. The story will be completed only when the goal is won. Even as these words are being written, American sailors are beleaguered in the frozen North, and others are preparing to follow them thither, so that the narrative here set forth must be accepted as only a partial story of a quest still being prosecuted.

In the private office of the President of the United States at Washington, stands a massive oaken desk. It has been a passive factor in the making of history, for at it have eight presidents sat, and papers involving almost the life of the nation, have received the executive signature upon its smooth surface. The very timbers of which it is built were concerned in the making of history of another sort, for they were part of the frame of the stout British ship "Resolute," which, after a long search in the Polar regions for the hapless Sir John Franklin—of whom more hereafter—was deserted by her crew in the Arctic pack, drifted twelve hundred miles in the ice, and was then discovered and brought back home as good as new by Captain Buddington of the stanch American whaler, "George and Henry." The sympathies of all civilized peoples, and particularly of English-speaking races, were at that time strongly stirred by the fate of Franklin and his brave companions, and so Congress appropriated $40,000 for the purchase of the vessel from the salvors, and her repair. Refitted throughout, she was sent to England and presented to the Queen in 1856. Years later, when broken up, the desk was made from her timbers and presented by order of Victoria to the President of the United States, who at that time was Rutherford B. Hayes. It stands now in the executive mansion, an enduring memorial of one of the romances of a long quest full of romance—the search for the North Pole.

In all ages, the minds of men of the exploring and colonizing nations, have turned toward the tropics as the region of fabulous wealth, the field for profitable adventure. "The wealth of the Ind," has passed into proverb. Though exploration has shown that, it is the flinty North that hides beneath its granite bosom the richest stores of mineral wealth, almost four centuries of failure and disappointment were needed to rid men's minds of the notion that the jungles and the tropical forests were the most abundant hiding-places of gold and precious stones. The wild beauty of the tropics, the cloudless skies, the tangled thickets, ever green and rustling with a restless animal life, the content and amiability of the natives, combined in a picture irresistibly attractive to the adventurer. Surely where there was so much beauty, so much of innocent joy in life, there must be the fountain of perpetual youth, there must be gold, and diamonds, and sapphires—all those gewgaws, the worship of which shows the lingering taint of barbarism in the civilized man, and for which the English, Spanish, and Portuguese adventurers of three centuries ago, were ready to sacrifice home and family, manhood, honor, and life.

So it happened that in the early days of maritime adventure the course of the hardy voyagers was toward the tropics, and they made of the Spanish Main a sea of blood, while Pizzarro and Cortez, and after them the dreaded buccaneers, sacked towns, betrayed, murdered, and outraged, destroyed an ancient civilization and fairly blotted out a people, all in the mad search for gold. Men only could have been guilty of such crimes, for man along, among animals endowed with life, kills for the mere lust of slaughter.

And yet, man alone stands ready to risk his life for an idea, to brave the most direful perils, to endure the most poignant suffering that the world's store of knowledge may be increased, that science may be advanced, that just one more fact may be added to the things actually known. If the record of man in the tropics has been stained by theft, rapine, and murder, the story of his long struggle with the Arctic ice, offers for his redemption a series of pictures of self-sacrifice, tenderness, honor, courage, and piety. No hope of profit drew the seamen of all maritime nations into the dismal and desolate ice-floes that guard the frozen North. No lust for gold impelled them to brave the darkness, the cold, and the terrifying silence of the six-months Arctic night. The men who have—thus far unsuccessfully—fought with ice-bound nature for access to the Pole, were impelled only by honorable emulation and scientific zeal.

The earlier Arctic explorers were not, it is true, searchers for the North Pole. That quest—which has written in its history as many tales of heroism, self-sacrifice, and patient resignation to adversity, as the poets have woven about the story of chivalry and the search for the Holy Grail—was begun only in the middle of the last century, and by an American. But for three hundred years English, Dutch, and Portuguese explorers, and the stout-hearted

American whalemen, had been pushing further and further into the frozen deep. The explorers sought the "Northwest Passage," or a water route around the northern end of North America, and so on to India and the riches of the East. Sir John Franklin, in the voyage that proved his last, demonstrated that such a passage could be made, but not for any practical or useful purpose. After him it was abandoned, and geographical research, and the struggle to reach the pole, became the motives that took men into the Arctic.

"But why," many people ask, with some reason, "should there be this determined search for the North Pole. What good will come to the world with its discovery? Is it worth while to go on year after year, pouring out treasure and risking human lives, merely that any hardy explorer may stand at an imaginary point on the earth's surface which is already fixed geographically by scientists?"

Let the scientists and the explorers answer, for to most of us the questions do not seem unreasonable.

Naturally, with the explorers' love for adventure, eagerness to see any impressive manifestations of nature's powers, and the ambition to attain a spot for which men have been striving for half a century, are the animating purposes. So we find Fridjof Nansen, who for a time held the record of having attained the "Furthest North," writing on this subject to an enquiring editor: "When man ceases to wish to know and to conquer every foot of the earth, which was given him to live upon and to rule, then will the decadence of the race begin. Of itself, that mathematical point which marks the northern termination of the axis of our earth, is of no more importance than any other point within the unknown polar area; but it is of much more importance that this particular point be reached, because there clings about it in the imagination of all mankind, such fascination that, till the Pole is discovered, all Arctic research must be affected, if not overshadowed, by the yearning to attain it."

George W. Melville, chief engineer of the United States Navy, who did such notable service in the Jeanette expedition of 1879, writes in words that stir the pulse:

"Is there a better school of heroic endeavor than the Arctic zone? It is something to stand where the foot of man has never trod. It is something to do that which has defied the energy of the race for the last twenty years. It is something to have the consciousness that you are adding your modicum of knowledge to the world's store. It is worth a year of the life of a man with a soul larger than a turnip, to see a real iceberg in all its majesty and grandeur. It is worth some sacrifice to be alone, just once, amid the awful silence of the Arctic snows, there to communicate with the God of nature, whom the thoughtful man finds best in solitude and silence, far from the haunts of men—alone with the Creator."

Thus the explorers. The scientists look less upon the picturesque and exciting side of Arctic exploration, and more upon its useful phases. "It helps to solve useful problems in the physics of the world," wrote Professor Todd of Amherst college. "The meteorology of the United States to-day; perfection of theories of the earth's magnetism, requisite in conducting surveys and navigating ships; the origin and development of terrestrial fauna and flora; secular variation of climate; behavior of ocean currents—all these are fields of practical investigation in which the phenomena of the Arctic and Antarctic worlds play a very significant role."

Lieutenant Maury, whose eminent services in mapping the ocean won him international honors, writes of the polar regions:

"There icebergs are launched and glaciers formed. There the tides have their cradle, the whales their nursery. There the winds complete their circuits, and the currents of the sea their round in the wonderful system of inter-oceanic circulation. There the aurora borealis is lighted up, and the trembling needle brought to rest, and there, too, in the mazes of that mystic circle, terrestrial forces of occult power, and vast influence upon the well-being of men, are continually at play.... Noble daring has made Arctic ice and waters classic ground. It is no feverish excitement nor vain ambition that leads man there. It is a higher feeling, a holier motive, a desire to look into the works of creation, to comprehend the economy of our planet, and to grow wiser and better by the knowledge."

Nor can it be said fairly that the polar regions have failed to repay, in actual financial profit, their persistent invasion by man. It is estimated by competent statisticians, that in the last two centuries no less than two thousand million dollars' worth of furs, fish, whale-oil, whalebone, and minerals, have been taken out of the ice-bound seas.

THEY FELL DOWN AND DIED AS THEY WALKED

The full story—at once sorrowful and stimulating—of Arctic exploration, can not be told here. That would require volumes rather than a single chapter. Even the part played in it by Americans can be sketched in outline only. But it is worth remembering that the systematic attack of our countrymen upon the Arctic fortress, began with an unselfish and humane incentive. In 1845 Sir John Franklin, a gallant English seaman, had set sail with two stout ships and 125 men, to seek the Northwest Passage. Thereafter no word was heard from him, until, years later, a searching party found a cairn of stones on a desolate, ice-bound headland, and in it a faintly written record, which told of the death of Sir John and twenty-four of his associates. We know now, that all who set out on this ill-fated expedition, perished. Struggling to the southward after abandoning their ships, they fell one by one, and their lives ebbed away on the cruel ice. "They fell down and died as they walked," said an old Esquimau woman to Lieutenant McClintock, of the British navy, who sought for tidings of them, and, indeed, her report found sorrowful verification in the skeletons discovered years afterward, lying face downward in the snow. To the last man they died. Think of the state of that last man—alone in the frozen wilderness! An eloquent writer, the correspondent McGahan, himself no stranger to Arctic pains and perils, has imagined that pitiful picture thus:

"One sees this man after the death of his last remaining companion, all alone in that terrible world, gazing round him in mute despair, the sole, living thing in that dark frozen universe. The sky is somber, the earth whitened with a glittering whiteness that chills the heart. His clothing is covered with frozen snow, his face lean and haggard, his beard a cluster of icicles. The setting sun looks back to see the last victim die. He meets her sinister gaze with a steady eye, as though bidding her defiance. For a few minutes they glare at each other, then the curtain is drawn, and all is dark."

As fears for Franklin's safety deepened into certainty of his loss with the passage of months and years, a multitude of searching expeditions were sent out, the earlier ones in the hope of rescuing him; the later ones with the purpose of discovering the records of his voyage, which all felt sure must have been cached at some accessible point. Americans took an active—almost a leading—part in these expeditions, braving in them the same perils which had overcome the stout English knight. By sea and by land they sought him. The story of the land expeditions, though full of interest, is foreign to the purpose of this work, and must be passed over with the mere note that Charles F. Hall, a Cincinnati journalist, in 1868-69, and Lieutenant Schwatka, and W.H. Gilder in 1878-79 fought their way northward to the path followed by the English explorer, found many relics of his expedition, and from the Esquimaux gathered indisputable evidence of his fate. By sea the United States was represented in the search for Franklin, by the ships "Advance" and "Rescue." They accomplished little of importance, but on the latter vessel was a young navy surgeon, Dr. Elisha Kent Kane, who was destined to make notable contributions to Arctic knowledge, both as explorer and writer.

One who studies the enormous volume of literature in which the Arctic story is told, scarcely can fail to be impressed by the pertinacity with which men, after one experience in the polar regions, return again and again to the quest for adventure and honors in the ice-bound zone. The subaltern on the expedition of to-day, has no sooner returned than he sets about organizing a new expedition, of which he may be commander. The commander goes into the ice time and again until, perhaps, the time comes when he does not come out. The leader of a rescue party becomes the leader of an exploring expedition, which in its turn, usually comes to need rescue.

So we find Dr. Kane, who was surgeon of an expedition for the rescue of Franklin, commanding four years later the brig "Advance," and voyaging northward through Baffin's Bay. Narrowly, indeed, he escaped the fate of the man in the search for whom he had gained his first Arctic experience. His ship, beset by ice, and sorely wounded, remained fixed and immovable for two years. At first the beleaguered men made sledge journeys in every direction for exploratory purposes, but the second year they sought rather by determined, though futile dashes across the rugged surface of the frozen sea, to find some place of refuge, some hope of emancipation from the thraldom of the ice. The second winter all of the brig except the hull, which served for shelter, was burned for fuel; two men had died, and many were sick of scurvy, the sledge dogs were all dead, and the end of the provisions was in sight. In May, 1855, a retreat in open boats, covering eighty-five days and over fifty miles of open sea, brought the survivors to safety.

When men have looked into the jaws of death, it might be thought they would strenuously avoid such another view. But there is an Arctic fever as well as an Arctic chill, and, once in the blood, it drags its victim irresistibly to the frozen North, until perhaps he lays his bones among the icebergs, cured of all fevers forever. And so, a year or two after the narrow escape of Dr. Kane, the surgeon of his expedition, Dr. Isaac I. Hayes, was hard at work fitting out an expedition of which he was to be commander, to return to Baffin's Bay and Smith sound, and if possible, fight its way into that open sea, which Dr. Hayes long contended surrounded the North Pole. No man in the Kane expedition had encountered greater perils, or withstood more cruel suffering than Dr. Hayes. A boat trip which he made in search of succor, has passed into Arctic history as one of the most desperate expedients ever adopted by starving men. But at the first opportunity he returned again to the scenes of his peril and his pain. His expedition, though conducted with spirit and determination, was not of great scientific value, as he was greatly handicapped in his observations by the death of his astronomer, who slipped through thin ice into the sea, and froze to death in his water-soaked garments.

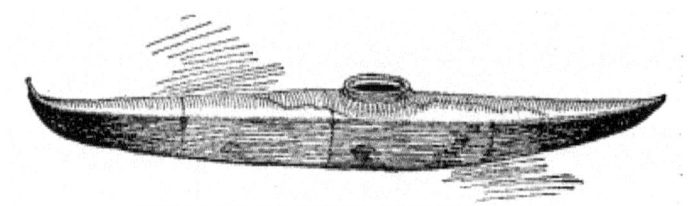

"THE TREACHEROUS KAYAK"

A most extraordinary record of daring and suffering in Arctic exploration was made by Charles F. Hall, to whom I have already referred. Beginning life as an engraver in Cincinnati, he became engrossed in the study of Arctic problems, as the result of reading the stories of the early navigators. Every book bearing on the subject in the library of his native city, was eagerly read, and his enthusiasm infected some of the wealthy citizens, who gathered for his use a very considerable collection of volumes. Mastering all the literature of the Arctic, he determined to undertake himself the arduous work of the explorer. Taking passage on a whaler, he spent several years among the Esquimaux, living in their crowded and fetid *igloos*, devouring the blubber and uncooked fish that form their staple articles of diet, wearing their garb of furs, learning to navigate the treacherous kayak in tossing seas, to direct the yelping, quarreling team of dogs over fields of ice as rugged as the edge of some monstrous saw, studying the geography so far as known of the Arctic regions, perfecting himself in all the arts by which man has contested the supremacy of that land with the ice-king. In 1870, with the assistance of the American Geographical Society, Hall induced the United States Government to fit him out an expedition to seek the North Pole—the first exploring party ever sent out with that definite purpose. The steamer "Polaris," a converted navy tug, which General Greely says was wholly unfit for Arctic service, was given him, and a scientific staff supplied by the Government, for though Hall had by painstaking endeavor qualified himself to lead an expedition, he had not enjoyed a scientific education. Neither was he a sailor like DeLong, nor a man trained to the command of men like Greely. Enthusiasm and natural fitness with him took the place of systematic training. But with him, as with so many others in this world, the attainment of the threshold of his ambition proved to be but opening the door to death. By a sledge journey from his ship he reached Cape Brevoort, above latitude 82, at that time the farthest north yet attained, but the exertion proved too much for him, and he had scarcely regained his ship when he died. His name will live, however, in the annals of the Arctic, for his contributions to geographical knowledge were many and precious.

THE SHIP WAS CAUGHT IN THE ICE PACK

The men who survived him determined to continue his work, and the next summer two fought their way northward a few miles beyond the point attained by Hall. But after this achievement the ship was caught in the ice-pack, and for two months drifted about, helpless in that unrelenting grasp. Out of this imprisonment the explorers escaped through a disaster, which for a time put all their lives in the gravest jeopardy, and the details of which seem almost incredible. In October, when the long twilight which precedes the polar night, had already set in, there came a fierce gale, accompanied by a tossing, roaring sea. The pack, racked by the surges, which now raised it with a mighty force, and then rolling on, left it to fall unsupported, began to go to pieces. The whistling wind accelerated its destruction, driving the floes far apart, heaping them up against the hull of the ship until the grinding and the prodigious pressure opened her seams and the water rushed in. The cry that the ship was sinking rung along the decks, and all hands turned with desperate energy to throwing out on the ice-floe to windward, sledges, provisions, arms, records—everything that could be saved against the sinking of the ship, which all thought was at hand. Nineteen of the ship's company were landed on the floe to carry the material away from its edge to a place of comparative safety. The peril seemed so imminent that the men in their panic performed prodigious feats of strength—lifting and handling alone huge boxes, which at ordinary times, would stagger two men. A driving, whirling snowstorm added to the gloom, confusion, and terror of the scene, shutting out almost completely those on the ice from the view of those still on the ship. In the midst of the work the cry was raised that the floes were parting, and with incredible rapidity the ice broke away from the ship on every side, so that communication between those on deck and those on the floe was instantly cut off by a broad interval of black and tossing water, while the dark and snow-laden air cut off vision on every side. The cries of those on the ice mingled with those from the fast vanishing ship, for each party thought itself in the more desperate case. The ice was fast going to pieces, and boats were plying in the lanes of water thus opened, picking up those clinging to smaller cakes of ice and transporting them to the main floe. On the ship the captain's call had summoned all hands to muster, and they gazed on each other in dumb despair as they saw how few of the ship's company remained. All were sent to the pumps, for the water in the hold was rising with ominous rapidity. The cry rang out that the steam-pumps must be started if the ship was to be saved, but long months had passed since any fire had blazed under those boilers, and to get up steam was a work of hours. With tar-soaked oakum and with dripping whale blubber the engineer strove to get the fires roaring, the while the men on deck toiled with desperate energy at the hand-pumps. But the water gained on them. The ship sunk lower and lower in the black ocean, until a glance over the side could tell all too plainly that she was going to her fate. Now the water begins to ooze through the cracks in the engine-room floor, and break in gentle ripples about the feet of the firemen. If it rises much higher it will flood the fire-boxes, and then all will be over, for there is not one boat left on the ship—all were landed on the now invisible floe. But just as all hope was lost there came a faint hissing of steam, the pumps began slowly moving, and then settled down into their monotonous "chug-chug," the sweetest sound, that day, those desperate mariners had ever heard. They were saved by the narrowest of chances.

ADRIFT ON AN ICE-FLOE

We must pass hastily to the sequel of this seemingly irreparable disaster. The "Polaris" was beached, winter quarters established, and those who had clung to the ship spent the winter building boats, in which, the following spring, they made their way southward until picked up by a whaler. Those on the floe drifted at the mercy of the wind and tide 195 days, making over 1300 miles to the southward. As the more temperate latitudes were reached, and the warmer days of spring came on, the floe began going to pieces, and they were continually confronted with the probability of being forced to their boat for safety—one boat, built to hold eight, and now the sole reliance of nineteen people. It is hard to picture through the imagination the awful strain that day and night rested upon the minds of these hapless castaways. Never could they drop off to sleep except in dread that during the night the ice on which they slept, might split, even under their very pallets, and they be awakened by the deathly plunge into the icy water. Day and night they were startled and affrighted by the thunderous rumblings and cracking of the breaking floe—a sound that an experienced Arctic explorer says is the most terrifying ever heard by man, having in it something of the hoarse rumble of heavy artillery, the sharp and murderous crackle of machine guns, and a kind of titanic grinding, for which there is no counterpart in the world of tumult. Living thus in constant dread of death, the little company drifted on, seemingly miraculously preserved. Their floe was at last reduced from a great sheet of ice, perhaps a mile or more square, to a scant ten yards by seventy-five, and this rapidly breaking up. In two days four whalers passed near enough for them to see, yet failed to see them, but finally their frantic signals attracted attention, and they were picked up—not only the original nineteen who had begun the drift six months earlier, but one new and helpless passenger, for one of the Esquimau women had given birth to a child while on the ice.

The next notable Arctic expedition from the United States had its beginning in journalistic enterprise. Mr. James Gordon Bennett, owner of the *New York Herald*, who had already manifested his interest in geographical work by sending Henry M. Stanley to find Livingston in the heart of the Dark Continent, fitted out the steam yacht "Pandora," which had already been used in Arctic service, and placed her at the disposal of Lieutenant DeLong, U.S.N., for an Arctic voyage. The name of the ship was changed to "Jeannette," and control of the expedition was vested in the United States Government, though Mr. Bennett's generosity defrayed all charges. The vessel was manned from the navy, and Engineer Melville, destined to bear a name great among Arctic men, together with two navy lieutenants, were assigned to her. The voyage planned was then unique among American Arctic expeditions, for instead of following the conventional route north through Baffin's Bay and Smith Sound, the "Jeannette" sailed from San Francisco and pushed northward through Bering Sea. In July, 1879, she weighed anchor. Two years after, no word having been heard of her meanwhile, the inevitable relief expedition was sent out—the steamer "Rodgers," which after making a gallant dash to a most northerly point, was caught in the ice-pack and there burned to the water's edge, her crew, with greatest difficulty, escaping, and reaching home without one ray of intelligence of DeLong's fate.

That fate was bitter indeed, a trial by cold, starvation, and death, fit to stand for awesomeness beside Greely's later sorrowful story. From the very outset evil fortune had attended the "Jeannette." Planning to winter on Wrangle Land—then thought to be a continent—DeLong caught in the ice-pack, was carried past its northern end, thus proving it to be an island, indeed, but making the discovery at heavy cost. Winter in the pack was attended with severe hardships and grave perils. Under the influence of the ocean currents and the tides, the ice was continually breaking up and shifting, and each time the ship was in imminent danger of being crushed. In his journal DeLong tries to describe the terrifying clamor of a shifting pack. "I know of no sound on shore that can be compared with it," he writes. "A rumble, a shriek, a groan, and the crash of a falling house all combined, might serve to convey an idea of the noise with which this motion of the ice-floe is accompanied. Great masses from fifteen to twenty-five feet in height, when up-ended, are sliding along at various angles of elevation and jam, and between and among them are large and confused masses of débris, like a marble yard adrift. Occasionally a stoppage occurs; some piece has caught against or under our floe; there follows a groaning and crackling, our floe bends and humps up in places like domes. Crash! The dome splits, another yard of floe edge breaks off, the pressure is relieved, and on goes again the flowing mass of rumbles, shrieks, groans, etc., for another spell."

DELONG'S MEN DRAGGING THEIR BOATS OVER THE ICE

Time and again this nerve-racking experience was encountered. More than once serious leaks were started in the ship, which had to be met by working the pumps and building false bulwarks in the hold; but by the exercise of every art known to sailors, she was kept afloat and tenable until June 11, 1881, when a fierce and unexpected nip broke her fairly in two, and she speedily sunk. There followed weeks and months of incessant and desperate struggling with sledge and boat against the forces of polar nature. The ship had sunk about 150 miles from what are known as the New Siberian Islands, for which DeLong then laid his course. The ice was rugged, covered with soft snow, which masked treacherous pitfalls, and full of chasms which had to be bridged. Five sleds and three boats were dragged by almost superhuman exertions, the sick feebly aiding the sturdy in the work. Imagine the disappointment, and despair of the leader, when, after a full week of this cruel labor, with provisions ever growing more scanty, an observation showed him they were actually twenty-eight miles further away from their destination than when they started! While they were toiling south, the ice-floe over which they were plodding was drifting more rapidly north. *Nil desperandum* must ever be the watchword of Arctic expeditions, and DeLong, saying nothing to the others of his discovery, changed slightly the course of his march and labored on. July 19 they reached an island hitherto unknown, which was thereupon named Bennett Island. A curious feature of the toilsome march across the ice, was that, though the temperature seldom rose to the freezing point, the men complained bitterly of the heat and suffered severely from sun-burn.

At Bennett Island they took to the boats, for now open water was everywhere visible. DeLong was making for the Lena River in Siberia, where there were known to be several settlements, but few of his party were destined to reach it. In a furious storm, on the 12th of September, the three boats were separated. One, commanded by Lieutenant Chipp, with eight men, must have foundered, for it was never again heard of. A second, commanded by George W. Melville, afterward chief engineer of the United States Navy, found one of the mouths of the Lena River, and ascending it reached a small Siberian village. Happy would it have been had DeLong and his men discovered the same pathway to safety, but the Lena is like our own Mississippi, a river with a broad delta and a multiplicity of mouths. Into an estuary, the banks of which were untrodden by man, and which itself was too shallow for navigation for any great distance, remorseless fate led DeLong. Forced soon to take to their sleds again, his companions toiled painfully along the river bank, with no known destination, but bearing ever to the south—the only way in which hope could possibly lie. Deserted huts and other signs of former human habitation were plenty, but nothing living crossed their path. At last, the food being at the point of exhaustion, and the men too weary and weak for rapid travel, DeLong chose two of the sturdiest,

Nindemann and Noros, and sent them ahead in the hope that they might find and return with succor. The rest stumbled on behind, well pleased if they could advance three miles daily. Food gave out, then strength. Resignation took the place of determination. DeLong's journal for the last week of life is inexpressibly pitiful:

"Sunday, October 23—133d Day: Everybody pretty weak. Slept or rested all day, and then managed to get in enough wood before dark. Read part of divine service. Suffering in our feet. No foot-gear.

"Monday, October 24—134th Day: A hard night.

"Tuesday, October 25—135th Day.

"Wednesday, October 26—136th day.

"Thursday, October 27—137th Day: Iverson broken down.

"Friday, October 28—138th Day: Iverson died during early morning.

"Saturday, October 29th—139th Day: Dressier died during the night.

"Sunday, October 30—140th Day: Boyd and Cortz died during the night. Mr. Collins dying."

This is the last entry. The hand that penned it, as the manuscript shows, was as firm and steady as though the writer were sitting in his library at home. Words are spelled out in full, punctuation carefully observed. How long after these words were set down DeLong too died, none may ever know; but when Melville, whom Nindemann and Noros had found after sore privations, reached the spot of the death camp, he came upon a sorrowful scene. "I came upon the bodies of three men partly buried in the snow," he writes, "one hand reaching out, with the left arm of the man reaching way above the surface of the snow—his whole left arm. I immediately recognized them as Captain DeLong, Dr. Ambler, and Ah Sam, the cook.... I found the journal about three or four feet in the rear of DeLong—that is, it looked as though he had been lying down, and with his left hand tossed the book over his shoulder to the rear, or to the eastward of him."

How these few words bring the whole scene up before us! Last, perhaps, of all to die, lying by the smoldering fire, the ashes of which were in the middle of the group of bodies when found, DeLong puts down the final words which tell of the obliteration of his party, tosses the book wearily over his shoulder, and turns on his side to die. And then the snow, falling gently, pitifully covers the rigid forms and holds them in its pure embrace until loyal friends seek them out, and tell to the world that again brave lives have been sacrificed to the ogre of the Arctic.

While DeLong and his gallant comrades of the United States Navy were dying slowly in the bleak desert of the Lena delta, another party of brave Americans were pushing their way into the Arctic circle on the Atlantic side of the North American continent. The story of that starvation camp in desolate Siberia was to be swiftly repeated on the shores of Smith Sound, and told this time with more pathetic detail, for of Greely's expedition, numbering twenty-five, seven were rescued after three years of Arctic suffering and starving, helpless, and within one day of death. They had seen their comrades die, destroyed by starvation and cold, and passing away in delirium, babbling of green fields and plenteous tables. From the doorway of the almost collapsed tent, in which the seven survivors were found, they could see the row of shallow graves in which their less fortunate comrades lay interred—all save two, whom they had been too weak to bury. No story of the Arctic which has come to us from the lips of survivors, has half the pathos, or a tithe of the pitiful interest, possessed by this story of Greely.

Studying to-day the history of the Greely expedition, it seems almost as if a malign fate had determined to bring disaster upon him. His task was not so arduous as a determined search for the Pole, or the Northwest Passage. He was ordered by the United States Government to establish an observation station on Lady Franklin Bay, and remain there two years, conducting, meanwhile, scientific observations, and pressing exploratory work with all possible zeal. The enterprise was part of a great international plan, by which each of the great nations was to establish and maintain such an observation station within the Arctic circle, while observations were to be carried on in all at once. The United States agreed to maintain two such stations, and the one at Point Barrow, north of Alaska, was established, maintained, and its tenants brought home at the end of the allotted time without disaster.

Greely was a lieutenant in the United States Army, and his expedition was under the immediate direction of the Secretary of War—at that time Robert Lincoln, son of the great war President. Some criticism was expressed at the time and, indeed, still lingers in the books of writers on the subject, concerning the fitness of an army officer to direct an Arctic voyage. But the purpose of the expedition was largely to collect scientific facts bear-on weather, currents of air and sea, the duration and extent of magnetic and electrical disturbances—in brief, data quite parallel to those which the United States signal service collects at home. So the Greely expedition was made an adjunct to the signal service, which in its turn is one of the bureaus of the War Department. Two army lieutenants, Lockwood and Klingsbury, and twenty men from the rank and file of the army and signal corps, were selected to form the party. An astronomer was needed, and Edward Israel, a young graduate of the University of Michigan, volunteered. George W. Rice volunteered as photographer. Both were enlisted in the army and given the rank of sergeant.

It is doubtful if any polar expedition was ever more circumstantially planned—none has resulted more disastrously, save Sir John Franklin's last voyage. The instructions of the War Department were as explicit as human foresight and a genius for detail could make them. Greely was to proceed to some point on Lady Franklin Bay, which enters the mainland of North America at about 81° 44' north latitude, build his station, and prepare for a two-years' stay. Provisions for three years were supplied him. At the end of one year it was promised, a relief ship should be sent him, which failing for any cause to reach the station, would cache supplies and dispatches at specified points. A year later a second relief ship would be sent to bring the party home, and if for any reason this ship should fail to make the station, then Greely was to break camp and sledge to the southward, following the east coast of the mainland, until he met the vessel, or reached the point at which fresh supplies were to be cached. No plan could have been better devised—none ever failed more utterly.

Arctic travel is an enigma, and it is an enigma never to be solved twice in the same way. Whalers, with the experience of a lifetime in the frozen waters, agree that the lessons of one voyage seldom prove infallible guides for the conduct of the next. Lieutenant Schwatka, a veteran Arctic explorer, said in an official document that the teachings of experience were often worse than useless in polar work. And so, though the Washington authorities planned for the safety of Greely according to the best guidance that the past could give them, their plans failed completely. The first relief ship did, indeed, land some stores—never, as the issue showed, to be reached by Greely—but the second expedition, composed of two ships, the "Proteus" and the "Yantic," accomplished nothing. The station was not reached, practically no supplies were landed, the "Proteus" was nipped by the ice and sunk, and the remnant of the expedition came supinely home, reporting utter failure. It is impossible to acquit the commanders of the two ships engaged in this abortive relief expedition of a lack of determination, a paucity of courage, complete incompetence. They simply left Greely to his fate while time still remained for his rescue, or at least for the convenient deposit of the vast store of provisions they brought home, leaving the abandoned explorers to starve.

The history of the Greely expedition and its achievements may well be sketched hastily, before the story of the catastrophe which overwhelmed it is told. As it was the most tragic of expeditions save one, Sir John Franklin's, so, too, it was the most fruitful in results, of any American expedition to the time of the writing of this book. Proceeding by the whaler "Proteus" in August, 1881, to the waters of the Arctic zone, Greely reached his destination with but little trouble, and built a commodious and comfortable station on the shores of Discovery Bay, which he called Fort Conger after a United States Senator from Michigan. A month remained before the Arctic night would set in, but the labor of building the house left little time for explorations, which were deferred until the following summer. Life at the station was not disagreeable. The house, stoutly built, withstood the bitter cold. Within there were books and games, and through the long winter night the officers beguiled the time with lectures and reading. Music was there, too, in impressive quantity, if not quality. "An organette with about fifty yards of music," writes Lieutenant Greely, "afforded much amusement, being particularly fascinating to our Esquimau, who never wearied grinding out one tune after another." The rigid routine of Arctic winter life was followed day by day, and the returning sun, after five months' absence, found the party in perfect health and buoyant spirits. The work of exploration on all sides began, the explorers being somewhat handicapped by the death of many of the sledge dogs from disease. Lieutenant Greely, Dr. Pavy, and Lieutenant Lockwood each led a party, but to the last named belong the honors, for he, with Sergeant Brainard and an Esquimau, made his way northward over ice that looked like a choppy sea suddenly frozen into the rigidity of granite, until he reached latitude 83° 24' north—the most northerly point then attained by any man—and still the record marking Arctic journey for an American explorer.

Winter came again under depressing circumstances. The first relief ship promised had not arrived, and the disappointment of the men deepened into apprehension lest the second, also, should fail them. Yet they went through the second winter in good health and unshaken morale, though one can not read such portions of Greely's diary as he has published, without seeing that the irritability and jealousy that seem to be the inevitable accompaniments of long imprisonment in an Arctic station, began to make their appearance. With the advent of spring the commander began to make his preparations for a retreat to the southward. If he had not then felt entire confidence in the promise of the War Department to relieve him without fail that summer, he would have begun his retreat early, and beyond doubt have brought all his men to safety before another winter set in or his provisions fell low. But as it was, he put off the start to the last moment, keeping up meanwhile the scientific work of the expedition, and sending out one party to cache supplies along the route of retreat. August 9, 1883, the march began—just two years after they had entered the frozen deep—Greely hoping to meet the relief ship oh the way. He did not know that three weeks before she had been nipped in the ice-pack, and sunk, and that her consort, the "Yantic," had gone impotently home, without even leaving food for the abandoned explorers. Over ice-fields and across icy and turbulent water, the party made its way for five hundred miles—four hundred miles of boating and one hundred of sledging—fifty-one days of heroic exertion that might well take the courage out of the stoutest heart. Sledging in the Arctic over "hummock" ice is, perhaps, the most wearing form of toil known to man, and with such heavy loads as Greely carried, every mile had to be gone over twice, and sometimes three times, as the men would be compelled to leave part of the load behind and go back after it. Yet the party was cheerful, singing and joking at their work, as one of the sergeants records. Finally they reached the vicinity of Cape Sabine, all in good health, with instruments and records saved, and with arms and ammunition enough to procure ample food in a land well stocked with game. But they did not worry very much about food, though their supply was by this time growing low. Was not Cape Sabine the spot at which the relief expeditions were to cache food, and could it be possible that the great United States Government would fail twice in an enterprise which any Yankee whaler would gladly take a contract to fulfill? And so the men looked upon the wilderness, and noted the coming on of the Arctic night again without fear, if with some disappointment. Less than forty days' rations remained. Eight months must elapse before any relief expedition could reach their camp, and far away in the United States the people were crying out in hot indignation that the authorities were basely leaving Greely and his devoted companions to their fate.

Pluckily the men set about preparing for the long winter. Three huts of stone and snow were planned, and while they were building, the hunters of the party scoured the neighboring ice-floes and pools for game—foxes, ptarmigan, and seals. There were no mistaken ideas concerning their deadly peril. Every man knew that if game failed, or if the provisions they hoped had been cached by the relief expeditions somewhere in the vicinity, could not be found, they might never leave that spot alive. Day by day the size of the rations was reduced. October 2 enough for thirty-five days remained, and at the request of the men, Greely so changed the ration as to provide for forty-five days. October 5 Lieutenant Lockwood noted in his diary:

"We have now three chances for our lives: First, finding American cache sufficient at Sabine or at Isabella; second, of crossing the straits when our present ration is gone; third, of shooting sufficient seal and walrus near by here to last during the winter."

How delusive the first chance proved we shall see later. The second was impractical, for the current carried the ice through the strait so fast, that any party trying to cross the floe, would have been carried south to where the strait widened out into Baffin's Bay before they could possibly pass the twenty-five miles which separated Cape Sabine from Littleton Island. Moreover, there was no considerable cache at the latter point, as Greely thought. As for the hunting, it proved a desperate chance, though it did save the lives of such of the party as were rescued. All feathered game took flight for the milder regions of the south when the night set in. The walrus which the hunters shot—two, Greely said, would have supplied food for all winter—and the seal sunk in almost every instance before the game could be secured.

The first, and most hopeful chance, was the discovery of cached provisions at Cape Sabine. To put this to the test, Rice, the photographer, who, though a civilian, proved to be one of the most determined and efficient men in the party, had already started for Sabine with Jens, the Esquimau. October 9 they returned, bringing the record of the sinking of the "Proteus," and the intelligence that there were about 1300 rations at, or near Cape Sabine. The record left at Cape Sabine by Garlington, the commander of the "Proteus" expedition, and which Rice brought back to the camp, read in part: "Depot landed ... 500 rations of bread, tea, and a lot of canned goods. Cache of 250 rations left by the English expedition of 1882 visited by me and found in good condition. Cache on Littleton Island. Boat at Isabella. U.S.S. 'Yantic' on way to Littleton Island with orders not to enter the ice. I will endeavor to communicate with these vessels at once.... Everything in the power of man will be done to rescue the (Greely's) brave men."

This discovery changed Greely's plans again. It was hopeless to attempt hauling the ten or twelve thousand pounds of material believed to be at Cape Sabine, to the site of the winter camp, now almost done, so Greely determined to desert that station and make for Cape Sabine, taking with him all the provisions and material he could drag. In a few days his party was again on the march across the frozen sea.

How inscrutable and imperative are the ways of fate! Looking backward now on the pitiful story of the Greely party, we see that the second relief expedition, intended to succor and to rescue these gallant men, was in fact the cause of their overwhelming disaster—and this not wholly because of errors committed in its direction, though they were many. When Greely abandoned the station at Fort Conger, he could have pressed straight to the southward without halt, and perhaps escaped with all his party—he could, indeed, have started earlier in the summer, and made escape for all certain. But he relied on the relief expedition, and held his ground until the last possible moment. Even after reaching Cape Sabine he might have taken to the boats and made his way southward to safety, for he says himself that open water was in sight; but the cheering news brought by Rice of a supply of provisions, and the promise left by Garlington, that all that men could do would be done for his rescue, led him to halt his journey at Cape Sabine, and go into winter quarters in the firm conviction that already another vessel was on the way to aid him. He did not know that Garlington had left but few provisions out of his great store, that the "Yantic" had fled without landing an ounce of food, and that the authorities at Washington had concluded that nothing more could be done that season—although whalers frequently entered the waters where Greely lay trapped, at a later date than that which saw the "Yantic's" precipitate retreat. Had he known these things, he says himself, "I should certainly have turned my back to Cape Sabine and starvation, to face a possible death on the perilous voyage along shore to the southward."

But not knowing them, he built a hut, and prepared to face the winter. It is worth noting, as evidence that Arctic hardships themselves, when not accompanied by a lack of food, are not unbearable, that at this time, after two years in the region of perpetual ice, the whole twenty-five men were well, and even cheerful. Depression and death came only when the food gave out.

The permanent camp, which for many of the party was to be a tomb, was fixed a few miles from Cape Sabine, by the side of a pool of fresh water—frozen, of course. Here a hut was built with stone walls three feet high, rafters made of oars with the blades cut off, and a canvas roof, except in the center, where an upturned whaleboat made a sort of a dome. Only under the whaleboat could a man get on his knees and hold himself erect; elsewhere the heads of the tall men touched the roof when they sat up in their sleeping bags on the dirt floor. With twenty-five men in sleeping bags, which they seldom left, two in each bag, packed around the sides of the hut, a stove fed with stearine burning in the center for the cooking of the insufficient food to which they were reduced, and all air from without excluded, the hut became a place as much of torture as of refuge.

The problem of food and the grim certainty of starvation were forced upon them with the very first examination of the caches of which Garlington had left such encouraging reports. At Cape Isabella only 144 pounds of meat was found, in Garlington's cache only 100 rations instead of 500 as he had promised. Moldy bread and dog biscuits fairly green with mold, though condemned by Greely, were seized by the famished men, and devoured ravenously without a thought of their unwholesomeness. When November 1 came, the daily ration for each man was fixed at six ounces of bread, four ounces of meat, and four ounces of vegetables—about a quarter of what would be moderate sustenance for a healthy man. By keeping the daily issue of food down to this pitiful amount Greely calculated that he would have enough to sustain life until the first of March, when with ten days' double rations still remaining, he would make an effort to cross the strait to Littleton Island, where he thought—mistakenly—that Lieutenant Garlington awaited him with ample stores. Of course all game shot added to the size of the rations, and that the necessary work of hunting might be prosecuted, the hunters were from the first given extra rations to maintain, their strength. Fuel, too, offered a serious problem. Alcohol, stearine, and broken wood from a whaleboat and barrels, were all employed. In order to get the greatest heat from the wood it was broken up into pieces not much larger than matches.

And yet packed into that noisome hovel, ill-fed and ill-clothed, with the Arctic wind roaring outside, the temperature within barely above freezing, and a wretched death staring each man in the face, these men were not without cheerfulness. Lying almost continually in their sleeping bags, they listened to one of their number reading aloud; such books as "Pickwick Papers," "A History of Our Own Times," and "Two on a Tower." Greely gave daily a lecture on geography of an hour or more; each man related, as best he could, the striking facts about his own State and city and, indeed, every device that ingenuity could suggest, was employed to divert their minds and wile away the lagging hours. Birthdays were celebrated by a little extra food—though toward the end a half a gill of rum for the celebrant, constituted the whole recognition of the day. The story of Christmas Day is inexpressibly touching as told in the simple language of Greely's diary:

"Our breakfast was a thin pea-soup, with seal blubber, and a small quantity of preserved potatoes. Later two cans of cloudberries were served to each mess, and at half-past one o'clock Long and Frederick commenced cooking dinner, which consisted of a seal stew, containing seal blubber, preserved potatoes and bread, flavored with pickled onions; then came a kind of rice pudding, with raisins, seal blubber, and condensed milk. Afterward we had chocolate, followed later by a kind of punch made of a gill of rum and a quarter of a lemon to each man.... Everybody was required to sing a song or tell a story, and pleasant conversation with the expression of kindly feelings, was kept up until midnight."

AN ARCTIC HOUSE

But that comparative plenty and good cheer did not last long. In a few weeks the unhappy men, or such as still clung to life, were living on a few shrimps, pieces of sealskin boots, lichens, and even more offensive food. The shortening of the ration, and the resulting hunger, broke down the moral sense of some, and by one device or another, food was stolen. Only two or three were guilty of this crime—an execrable one in such an emergency—and one of these, Private Henry, was shot by order of Lieutenant Greely toward the end of the winter. Even before Christmas, casualties which would have been avoided, had the party been well-nourished and strong, began. Ellison, in making a gallant dash for the cache at Isabella, was overcome by cold and fatigue, and froze both his hands and feet so that in time they dropped off. Only the tender care of Frederick, who was with him, and the swift rush of Lockwood and Brainard to his aid, saved him from death. It tells a fine story of the unselfish devotion of the men, that this poor wreck, maimed and helpless, so that he had to be fed, and incapable of performing one act in his own service, should have been nursed throughout the winter, fed with double portions, and actually saved living until the rescue party arrived, while many of those who cared for him yielded up their lives. The first to die was Cross, of scurvy and starvation, and he was buried in a shallow grave near the hut, all hands save Ellison turning out to honor his memory. Though the others clung to life with amazing tenacity, illness began to make inroads upon them, the gallant Lockwood, for example, spending weeks in Greely's sleeping bag, his mind wandering, his body utterly exhausted. But it was April before the second death occurred—one of the Esquimaux. "Action of water on the heart caused by insufficient nutrition," was the doctor's verdict—in a word, but a word all dreaded to hear, starvation.

Thereafter the men went fast. In a day or two Christiansen, an Esquimau, died. Rice, the sharer of his sleeping bag, was forced to spend a night enveloped in a bag with the dead body. The next day he started on a sledging trip to seek some beef cached by the English years earlier. Before the errand was completed, he, too, died, freezing to death in the arms of his companion, Frederick, who held him tenderly until the last, and stripped himself to the shirtsleeves in the icy blast, to warm his dying comrade. Then Lockwood died—the hero of the Farthest North; then Jewell. Jens, the untiring Esquimau hunter, was drowned, his kayak being cut by the sharp edge of a piece of ice. Ellis, Whisler, Israel, the astronomer, and Dr. Pavy, the surgeon, one by one, passed away.

But why continue the pitiful chronicle? To tell the story in detail is impossible here—to tell it baldly and hurriedly, means to omit from it all that makes the narrative of the last days of the Greely expedition worth reading; the unflagging courage of most of the men, the high sense of honor that characterized them, the tenderness shown to the sick and helpless, the pluck and endurance of Long and Brainard, the fierce determination of Greely, that come what might, the records of his expedition should be saved, and its honor bequeathed unblemished to the world. And so through suffering and death, despairing perhaps, but never neglecting through cowardice or lethargy, any expedient for winning the fight against death, the party, daily growing smaller, fought its way on through winter and spring, until that memorable day in June, when Colwell cut open the tent and saw, as the first act of the rescued sufferers, two haggard, weak, and starving men pouring all that was left of the brandy, down the throat of one a shade more haggard and weak than they.

Men of English lineage are fond of telling the story of the meeting of Stanley and Dr. Livingston in the depths of the African jungle. For years Livingston had disappeared from the civilized world. Everywhere apprehension was felt lest he had fallen a victim to the ferocity of the savages, or to the pestilential climate. The world rung with speculations concerning his fate. Stanley, commissioned to solve the mystery, by the same America journalist who sent DeLong into the Arctic, had cut his path through the savages and the jungle, until at the door of a hut in a clearing, he saw a white man who could be none but him whom he sought, for in all that dark and gloomy forest there was none other of white skin. Then Anglo-Saxon stolidity asserted itself. Men of Latin race would have rushed into each others' arms with loud rejoicings. Not so these twain.

"Dr. Livingston, I believe," said the newcomer, with the air of greeting an acquaintance on Fifth Avenue. "I am Mr. Stanley."

"I am glad to see you," was the response, and it might have taken place in a drawing-room for all the emotion shown by either man.

AN ESQUIMAU

That was a dramatic meeting in the tropical jungles, but history will not give second place to the encounter of the advance guard of the Greely relief expedition with the men they sought. The story is told with dramatic directness in Commander (now Admiral) Schley's book, "The Rescue of Greely."

"It was half-past eight in the evening as the cutter steamed around the rocky bluff of Cape Sabine, and made her way to the cove, four miles further on, which Colwell remembered so well.... The storm which had been raging with only slight intervals since early the day before, still kept up, and the wind was driving in bitter gusts through the opening in the ridge that followed the coast to the westward. Although the sky was overcast it was broad daylight—the daylight of a dull winter afternoon.... At last the boat arrived at the site of the wreck cache, and the shore was eagerly scanned, but nothing could be seen. Rounding the next point, the cutter opened out the cove beyond. There on the top of a little ridge, fifty or sixty yards above the ice-foot, was plainly outlined the figure of a man. Instantly the coxswain caught up his boathook and waved his flag. The man on the ridge had seen them, for he stooped, picked up a signal flag, and waved it in reply. Then he was seen coming slowly and cautiously down the steep rocky slope. Twice he fell down before he reached the foot. As he approached, still walking slowly and with difficulty, Colwell hailed him from the bow of the boat.

"'Who all are there left?'

"'Seven left.'

"As the cutter struck the ice Colwell jumped off, and went up to him. He was a ghastly sight. His cheeks were hollow, his eyes wild, his hair and beard long and matted. His army blouse, covering several thicknesses of shirts and jackets, was ragged and dirty. He wore a little fur cap and rough moccasins of untanned leather tied around the leg. As he spoke his utterance was thick and mumbling, and in his agitation his jaws worked in convulsive twitches. As the two met, the man, with a sudden impulse, took off his gloves and shook Colwell's hand.

"'Where are they?' asked Colwell, briefly.

"'In the tent,' said the man, pointing over his shoulder, 'over the hill—the tent's down.'

"'Is Mr. Greely alive?'

"'Yes, Greely's alive.'

"'Any other officers?'

"'No.' Then he repeated absently, 'The tent's down.'

"'Who are you?'

"'Long.'

"Before this colloquy was over Lowe and Norman had started up the hill. Hastily filling his pockets with bread, and taking the two cans of pemmican, Colwell told the coxswain to take Long into the cutter, and started after the others with Ash. Reaching the crest of the ridge and looking southward, they saw spread out before them a desolate expanse of rocky ground, sloping gradually from a ridge on the east to the ice-bound shore, which on the west made in and formed a cove. Back of the level space was a range of hills rising up eight hundred feet with a precipitous face, broken in two by a gorge, through which the wind was blowing furiously. On a little elevation directly in front was the tent. Hurrying on across the intervening hollow, Colwell came up with Lowe and Norman just as they were greeting a soldierly-looking man who had come out of the tent.

"As Colwell approached, Norman was saying to the man: 'There is the Lieutenant.'

"And he added to Lieutenant Colwell:

"'This is Sergeant Brainard.'

"Brainard immediately drew himself up to the position of the soldier, and was about to salute, when Colwell took his hand.

"At this moment there was a confused murmur within the tent, and a voice said: 'Who's there?'

"Norman answered, 'It's Norman—Norman who was in the "Proteus."'

"This was followed by cries of 'Oh, it's Norman,' and a sound like a feeble cheer.

"Meanwhile one of the relief party, who in his agitation and excitement was crying like a child, was down on his knees trying to roll away the stones that held the flapping tent-cloth.... Colwell called for a knife, cut a slit in the tent-cover, and looked in. It was a sight horror. On one side, close to the opening, with his face toward the opening, lay what was apparently a dead man. His jaw had dropped, his eyes were open, but fixed and glassy, his limbs were motionless. On the opposite side was a poor fellow, alive to be sure, but without hands or feet, and with a spoon tied to the stump of his right arm. Two others, seated on the ground in the middle, had just got down a rubber bottle that hung on the tent pole, and were pouring from it into a tin can. Directly opposite, on his hands and knees, was a dark man, with a long matted beard, in a dirty and tattered dressing-gown, with a little red tattered skull-cap on his head, and brilliant, staring eyes. As Colwell appeared he raised himself a little and put on a pair of eye-glasses.

"'Who are you?' asked Colwell.

"The man made no reply, staring at him vacantly.

"'Who are you?' again.

"One of the men spoke up. 'That's the Major—Major Greely."

"Colwell crawled in and took him by the hand, saying: 'Greely, is this you?'

"'Yes,' said Greely in a faint voice, hesitating and shuffling with his words, 'yes—seven of us left—here we are—dying—like men. Did what I came to do—beat the best record.'

"Then he fell back exhausted."

Slowly and cautiously the men were nursed back to life and health—all save poor Ellison, whose enfeebled constitution could not stand the shock of the necessary amputation of his mutilated limbs. The nine bodies buried in the shallow graves were exhumed and taken to the ship, Private Henry's body being found lying where it fell at the moment of his execution. At that time the castaways were too feeble to give even hasty sepulture to their dead. A horrible circumstance, reported by Commander Schley himself, was that the flesh of many of the bodies was cut from the bones—by whom, and for what end of cannibalism, can only be conjectured.

Following the disaster to the Greely expedition, came a period of lethargy in polar exploration, and when the work was taken up again, it was in ways foreign to the purpose of this book. Foreigners for a time led in activity, and in 1895 Fridjof Nansen in his drifting ship, the "Fram," attained the then farthest North, latitude 86° 14', while Rudolph Andree, in 1897, put to the test the desperate expedient of setting out for the Pole in a balloon from Dane's Island, Spitzbergen; but the wind that bore him swiftly out of sight, has never brought back again tidings of his achievement or his fate. Nansen's laurels were wrested from him in 1900 by the Duke of Abruzzi, who reached 86° 33' north. The stories of these brave men are fascinating and instructive, but they are no part of the story of the American sailor. Indeed, the sailor is losing his importance as an explorer in the Arctic. It has become clear enough to all that it is not to be a struggle between stout ships and crushing ice, but rather a test of the endurance of men and dogs, pushing forward over solid floes of heaped and corrugated ice, toward the long-sought goal. Two Americans in late years have made substantial progress toward the conquest of the polar regions. Mr. Walter Wellman, an eminent journalist, has made two efforts to reach the Pole, but met with ill-luck and disaster in each, though in the first he attained to latitude 81° to the northeast of Spitzbergen, and in the second he discovered and named many new islands about Franz Josef Land. Most pertinacious of all the American explorers, however, has been Lieutenant Robert E. Peary, U.S.N., who since 1886, has been going into the frozen regions whenever the opportunity offered—and when none offered he made one. His services in exploration and in mapping out the land and seas to the north of Greenland have been of the greatest value to geographical science, and at the moment of writing this book he is wintering at Cape Sabine, where the Greely survivors were found, awaiting the coming of summer to make a desperate dash for the goal, sought for a century, but still secure in its wintry fortifications, the geographical Pole. Nor is he wholly alone, either in his ambition or his patience. Evelyn B. Baldwin, a native of Illinois, with an expedition equipped by William Zeigler, of New York, and made up of Americans, is wintering at Alger Island, near Franz Josef Land, awaiting the return of the sun to press on to the northward. It is within the bounds of possibility that before this volume is fairly in the hands of its readers, the fight may be won and the Stars and Stripes wave over that mysterious spot that has awakened the imagination and stimulated the daring of brave men of all nations.

CHAPTER VII.

THE GREAT LAKES—THEIR SHARE IN THE MARITIME TRAFFIC OF THE UNITED STATES—THE EARLIEST RECORDED VOYAGERS—INDIANS AND FUR TRADERS—THE PIGMY CANAL AT THE SAULT STE. MARIE—BEGINNINGS OF NAVIGATION BY SAILS—DE LA SALLE AND THE "GRIFFIN"—RECOLLECTIONS OF EARLY LAKE SEAMEN—THE LAKES AS A HIGHWAY FOR WESTWARD EMIGRATION—THE FIRST STEAMBOAT—EFFECT OF MINERAL DISCOVERIES ON LAKE SUPERIOR—THE ORE-CARRYING FLEET—THE WHALEBACKS—THE SEAMEN OF THE LAKES—THE GREAT CANAL AT THE "SOO"—THE CHANNEL TO BUFFALO—BARRED OUT FROM THE OCEAN.

In the heart of the North American Continent, forming in part the boundary line between the United States and the British possessions to the north, lies that chain of great freshwater lakes bordered by busy and rapidly growing commonwealths, washing the water-fronts of rich and populous cities, and bearing upon their steely blue bosoms a commerce which outdoes that of the Mediterranean in the days of its greatest glory. The old salt, the able seaman who has rounded the Horn, the skipper who has stood unflinchingly at the helm while the green seas towered over the stern, looks with contempt upon the fresh-water sailor and his craft. Not so the man of business or the statesman. The growth of lake traffic has been one of the most marvelous and the most influential factors in the industrial development of the United States. By it has been systematized and brought to the highest form of organization the most economical form of freight carriage in the world. Through it has been made possible the enormous reduction in the price of American steel that has enabled us to invade foreign markets, and promises to so reduce the cost of our ships, that we may be able to compete again in ship-building, with the yards of the Clyde and the Tyne. Along the shores of these unsalted seas, great shipyards are springing up, that already build ships more cheaply than can be done anywhere else in the world, and despite the obstacles of shallow canals, and the treacherous channels of the St. Lawrence, have been able to build and send to tidewater, ocean ships in competition with the seacoast builders. The present of the lake marine is secure; its future is full of promise. Its story, if lacking in the elements of romance that attend upon the ocean's story, is well worth telling.

A decade more than two centuries ago a band of Iroquois Indians made their way in bark canoes from Lake Ontario up Lake Erie to the Detroit River, across Lake St. Clair, and thence through Lake Huron to Point Iroquois. They were the first navigators of the Great Lakes, and that they were not peace-loving boatmen, is certain from the fact that they traveled all these miles of primeval waterway for the express purpose of battle. History records that they had no difficulty in bringing on a combat with the Illinois tribes, and in an attempt to displace the latter from Point Iroquois, the invaders were destroyed after a six-days' battle.

It is still a matter of debate among philosophical historians, whether war, trade, or missionary effort has done the more toward opening the strange, wild places of the world. Each, doubtless, has done its part, but we shall find in the story of the Great Lakes, that the war canoes of the savages were followed by the Jesuit missionaries, and these in turn by the bateaux of the voyageurs employed by the Hudson Bay Company.

After the Iroquois had learned the way, trips of war canoes up and down the lakes, were annual occurrences, and warfare was almost perpetual. In 1680 the Iroquois, 700 strong, invaded Illinois, killed 1200 of the tribe there established, and drove the rest beyond the Mississippi. For years after the Iroquois nation were the rulers of the water-front between Lake Erie and Lake Huron. While this tribe was in undisputed possession, commerce had little to do with the navigation of the Great Lakes. The Indians went up and down the shores on long hunting trips, but war was the principal business, and every canoe was equipped for a fray at any time.

A story is told of a great naval battle that was fought on Lake Erie, nearly two centuries before the first steamer made its appearance on that placid water. A Wyandot prince, so the tale goes, fell in love with a beautiful princess of the Seneca tribe, who was the promised bride of a chief of her own nation. The warrior failed to win the heart of the dusky maiden, and goaded to desperation, entered the Senecas country by night, and carried off the lady. War immediately followed, and was prosecuted with great cruelty and slaughter for a long time. At last a final battle was fought, in which the Wyandots were worsted and forced to flee in great haste. The fugitives planned to cross the ice of the Straits (Detroit) River, but found it broken up and floating down stream. Their only alternative was to throw themselves on the floating ice and leap from cake to cake; they thus made their escape to the Canadian shore, and joined the tribes of the Pottawatomies, Ottawas, and Chippewas. A year later the Wyandots, equipped with light birch canoes, set out to defeat the Senecas, and succeeded in inducing them to give combat on the water. The Senecas made a fatal mistake and came out to meet the enemy in their clumsily-constructed boats hollowed out of the trunks of trees. After much maneuvering the birch canoe fleet proceeded down Lake Erie to the head of Long Point, with the Senecas in hot pursuit. In the center of the lake the Wyandots turned and gave the Senecas so hot a reception that they were forced to flee, but could not make good their escape in their clumsy craft, and were all slain but one man, who was allowed to return and report the catastrophe to his own nation. This closed the war.

Legends are preserved that lead to the belief that there may have been navigators of the Great Lakes before the Indians, and it is generally believed that the latter were not the first occupants of the Lake Superior region. It is said that the Lake Superior country was frequently visited by a barbaric race, for the purpose of obtaining copper, and it is quite possible that these people may have been skilled navigators.

THE WOODEN BATEAUX OF THE FUR TRADERS

Commercial navigation of the Great Lakes, curiously enough, first assumed importance in the least accessible portion. The Hudson Bay Company, always extending its territory toward the northwest, sent its bateaux and canoes into Lake Superior early in the seventeenth century. To accommodate this traffic the company dug a canal around the falls of the St. Marie River, at the point we now call "the Soo." In time this pigmy progenitor of the busiest canal in the world, became filled with débris, and its very existence forgotten; but some years ago a student in the thriving town of Sault Ste. Marie, poring over some old books of the Hudson Bay Company, noticed several references to the company's canal. What canal could it be? His curiosity was aroused, and with the aid of the United States engineers in charge of the new improvements, he began a painstaking investigation. In time the line of the old ditch was discovered, and, indeed, it was no more than a ditch, two and a half feet deep, by eight or nine wide. One lock was built, thirty-eight feet long, with a lift of nine feet. The floor and sills of this lock were discovered, and the United States Government has since rebuilt it in stone, that visitors to the Soo may turn from the massive new locks, through which steel steamships of eight thousand tons pass all day long through the summer months, to gaze on the strait and narrow gate which once opened the way for all the commerce of Lake Superior. But through that gate there passed a picturesque and historic procession. Canoes spurred along by tufted Indians with black-robed Jesuit missionaries for passengers; the wooden bateaux of the fur traders, built of wood and propelled by oars, and carrying gangs of turbulent trappers and voyageurs; the company's chief factors in swift private craft, making for the west to extend the influence of the great corporation still further into the wilderness, all passed through the little canal and avoided the roaring waters of the Ste. Marie. It was but a narrow gate, but it played its part in the opening of the West.

War, which is responsible for most of the checks to civilization, whether or not it may in some instances advance the skirmish line of civilized peoples, destroyed the pioneer canal. For in 1812 some Americans being in that part of the country, thought it would be a helpful contribution to their national defense if they blew up the lock and shattered the canal, as it was on Canadian soil. Accordingly this was done, of course without the slightest effect on the conflict then raging, but much to the discomfort and loss of the honest voyageurs and trappers of the Lake Superior region, whose interest in the war could hardly have been very serious.

So far as history records the first sailing vessel to spread its wings on the Great Lakes beyond Niagara Falls, was the "Griffin," built by the Chevalier de la Salle in 1679, near the point where Buffalo now stands. La Salle had brought to this point French ship-builders and carpenters, together with sailors, to navigate the craft when completed. It was his purpose to proceed in this vessel to the farthest corners of the Great Lakes, establish trading and trapping stations, and take possession of the country in the name of France. He was himself conciliatory with the Indians and liked by them, but jealousies among the French themselves, stirred up savage antagonism to him, and his ship narrowly escaped burning while still on the stocks. In August of 1679, however, she was launched, a brigantine of sixty tons burden, mounting five small cannon and three arquebuses. Her model is said to have been not unlike that of the caravels in which Columbus made his famous voyage, and copies of which were exhibited at the Columbian Exposition. Bow and stern were high and almost alike. Yet in this clumsy craft La Salle voyaged the whole length of Lake Erie, passed through the Detroit River, and St. Clair River and lake; proceeded north to Mackinaw, and thence south in Lake Michigan and into Green Bay. It was the first time any vessel under sail had entered those waters. Maps and charts there were none. The swift rushing waters of the Detroit River flowed smoothly over limestone reefs, which the steamers of to-day pass cautiously, despite the Government channels, cut deep and plainly lighted. The flats, that broad expanse of marsh permeated by a maze of false channels above Detroit, had to be threaded with no chart or guide. Yet the "Griffin" made St. Ignace in twenty days from having set sail, a record which is often not equaled by lumber schooners of the present time. From Green Bay, La Salle sent the vessel back with a cargo of furs that would have made him rich for life, had it ever reached a market. But the vessel disappeared, and for years nothing was heard of her. Finally La Salle learned that a half-breed pilot, who had shown signs of treachery on the outward trip, had persuaded the crew to run her ashore in the Detroit River, and themselves to take the valuable cargo. But the traitors had reckoned without the savage Indians of the neighborhood, who also coveted the furs and pelts. While the crew were trying to dispose of these the red men set upon them and slew them all. The "Griffin" never again floated on the lakes.

It is difficult to determine the time when sailing vessels next appeared upon the lakes, but it was certainly not for nearly seventy-five years. Captain Jonathan Carver reported a French schooner on Lake Superior about 1766, and in 1772 Alexander Harvey built a forty-ton sloop on the same lake, in which he sought the site of a famous copper mine. But it was long before Lake Superior showed more than an infrequent sail, though on Lake Erie small vessels soon became common. Even in 1820 the furs of Lake Superior were sent down to Chicago in bateaux.

Two small sailing vessels, the "Beaver" and the "Gladwin," which proved very valuable to the besieged garrison at Detroit in 1763, were the next sailing vessels on the lakes, and are supposed to have been built by the English the year previous. It is said, that through the refusal of her captain to take ballast aboard, the "Gladwin" was capsized on Lake Erie and lost, and the entire crew drowned. The "Royal Charlotte," the "Boston," and the "Victory" appeared on the lakes a few years later, and went into commission between Fort Erie (Buffalo) and Detroit, carrying the first year 1,464 bales of fur to Fort Erie, and practically establishing commercial navigation.

It is hard to look clearly into the future. If the recommendations of one J. Collins, deputy surveyor-general of the British Government, had governed the destiny of the Great Lakes, the traffic between Buffalo and the Soo by water, would to-day be in boats of fifteen tons or less. Under orders of the English Government, Collins in 1788 made a survey of all the lakes and harbors from Kingston to Mackinac, and in his report, expressing his views as to the size of vessels that should be built for service on the lakes, he said he thought that for service on Lake Ontario vessels should be seventy-five or eighty tons burden, and on Lake Erie, if expected to run to Lake Huron, they should be not more than fifteen tons. What a stretch of imagination is necessary to conceive of the great volume of traffic of the present time, passing Detroit in little schooners not much larger than catboats that skim around the lakes! Imagine such a corporation as the Northern Steamship Company, with its big fleet of steel steamers, attempting to handle its freight business in sailing vessels of a size that the average wharf-rat of the present time would disdain to pilot. What a rush of business there would be at the Marine Post-Office in Detroit, if some day this company would decide to cut off three of its large steamers and send out enough schooners of the size recommended by the English officer, to take their place! The fleet would comprise at least 318 vessels, and would require not fewer than 1500 seamen to navigate. It is sometimes said that there is a continual panorama of vessels passing up and down the rivers of the Great Lakes, but what if the Englishman had guessed right? Happily he did not, and vessels of 1500 tons can navigate the connecting waters of Lake Huron and Lake Erie much better than those of fifteen tons could in his time. That the early ship-builders did not pay much attention to J. Collins, is evident from the fact that, when the Detroit was surrendered to the Americans in 1796, twelve merchant vessels were owned there of from fifty to one hundred tons each.

"THE RED MEN SET UPON THEM AND SLEW THEM ALL"

At the close of the eighteenth century the American sailor had hardly superseded the red men as a navigator, and lake vessels were not much more plentiful than airships are nowadays. Indeed, the entire fleet in 1799, so far as can be learned, was as follows: The schooners "Nancy," "Swan," and

"Naegel;" the sloops "Sagina," "Detroit," "Beaver," "Industry," "Speedwell," and "Arabaska." This was the fleet, complete, of Lakes Huron, Erie, and Michigan.

"A wild-looking set were the first white sailors of the lakes," says Hubbard in his "Memorials of Half a Century." "Their weirdness was often enhanced by the dash of Indian blood, and they are better described as rangers of the woods and waters. Picturesque, too, they were in their red flannel or leather shirts and cloth caps of some gay color, finished to a point which hung over on one side with a depending tassel. They had a genuine love for their occupation, and muscles that never seemed to tire at the paddle and oar. These were not the men who wanted steamboats and fast sailing vessels. These men had a real love for canoeing, and from dawn to sunset, with only a short interval, and sometimes no midday rest, they would ply the oars, causing the canoe or barge to shoot through the water like a thing of life, but often contending against head winds and gaining little progress in a day's rowing."

ONE OF THE FIRST LAKE SAILORS

One of the earliest American sailors on a lake ship bigger than a bateau, was "Uncle Dacy" Johnson, of Cleveland, who sailed for fifty years, beginning about 1850. "When I was a chunk of a boy," says the old Captain in a letter to a New York paper, "I put a thirty-two pound bundle on my back and started on foot to Buffalo. I made the journey to Albany, N.Y., from Bridgeport, Conn., in sixteen days, which was nothing remarkable, as I had $3 in money, and a bundle of food. Many a poor fellow I knew started on the same journey with nothing but an axe. When I arrived at Buffalo I found a very small town—Cleveland, Sandusky, and Erie, were all larger. There were only two lighthouses on the lakes, one at Buffalo, which was the first one built, and the other one at Erie. Buffalo was then called Fort Erie, and was a struggling little town. My first trip as a sailor was made from Buffalo to Erie, which was then considered quite a voyage. From Buffalo to Detroit was looked upon as a long voyage, and a vessel of thirty-two tons was the largest ship on the lakes. In 1813 I was one of a crew of four who left Buffalo on the sloop 'Commencement' with a cargo of whisky for Erie. While beating along shore the English frigate 'Charlotte' captured us and two boatloads of red-coats boarded our vessel and took us prisoners. We were paroled on shipboard the same day, and before night concocted a scheme to get the Englishmen drunk on our whisky. One of our fellows got drunk first, and told of our intentions, the plot was frustrated, and we narrowly escaped being hung."

"TWO BOAT-LOADS OF REDCOATS BOARDED US AND TOOK US PRISONERS"

Once begun, the conquest of the lakes as a highway for trade was rapid. We who live in the days of railroads can hardly appreciate how tremendous was the impetus given to the upbuilding of a region if it possessed practicable waterways. The whole history of the settlement of the Middle West is told in the story of its rivers and lakes. The tide of immigration, avoiding the dense forests haunted by Indians, the rugged mountains, and the broad prairies into which the wheel of the heavy-laden wagon cut deep, followed the course of the Potomac and the Ohio, the Hudson, Mohawk, and the Great Lakes. Streams that have long since ceased to be thought navigable for a boy's canoe were made to carry the settlers' few household goods heaped on a flatboat. The flood of families going West created a demand that soon covered the lakes with schooners and brigs. Landed on the lake shore near some little stream, the immigrants would build flatboats, and painfully pole their way into the interior to some spot that took their fancy. Ohio, Indiana, Michigan, and Illinois thus filled up, towns growing by the side of streams now used only to turn mill-wheels, but which in their day determined where the prosperous settlement should be.

The steamboat was not slow in making its appearance on the lakes. In 1818, while it was still an experiment on the seaboard, one of these craft appeared on Lake Erie. The "Walk-in-the-Water" was her name, suggestive of Indian nomenclature and, withal, exceedingly descriptive. She made the trip from Buffalo to Detroit, not infrequently taking thirteen days. She was a side-wheeler, a model which still holds favor on the lower lakes, though virtually abandoned on the ocean and on Lake Superior. An oil painting of this little craft, still preserved, shows her without a pilot-house, steered by a curious tiller at the stern, with a smokestack like six lengths of stovepipe, and huge unboxed wheels. She is said to have been a profitable craft, often carrying as many as fifty passengers on the voyage, for which eighteen dollars was charged. For four years she held a monopoly of the business. Probably the efforts of Fulton and Livingstone to protect the monopoly which had been granted them by the State of New York, and the determination of James Roosevelt to maintain what he claimed to be his exclusive right to the vertical paddle-wheel, delayed the extension of steam navigation on the lakes as it did on the great rivers. After four years of solitary service on Lake Erie, the "Walk-in-the-Water" was wrecked in an October storm. Crowded with passengers, she rode out a heavy gale through a long night. At daybreak the cables parted and she went ashore, but no lives were lost. Her loss was considered an irreparable calamity by the settlers at the western end of the lake. "This accident," wrote an eminent citizen of Detroit, "may be considered one of the greatest misfortunes which has ever befallen Michigan, for, in addition to its having deprived us of all certain and speedy communication with the civilized world, I am fearful it will greatly check the progress of immigration and improvement."

It is scarcely necessary to note now that the apprehensions of the worthy citizen of Michigan were unfounded. Steam navigation on the lakes was no more killed by the loss of the pioneer craft than was transatlantic steam navigation ended by the disapproving verdict of the scientists. Nowhere in the world is there such a spectacle of maritime activity, nowhere such a continuous procession of busy cargo-ships as in the Detroit River, and through the colossal locks of the "Soo" canals. In 1827 the first steamboat reached the Sault Ste. Marie, bearing among her passengers General Winfield Scott, on a visit of inspection to the military post there, but she made no effort to enter the great lake. About five years later, the first "smoke boat," as the Indians called the steamers, reached Chicago, the pigmy forerunner of the fleet of huge leviathans that all the summer long, nowadays, blacken

Chicago's sky with their torrents of smoke, and keep the hurrying citizens fuming at the open draw of a bridge. All side-wheelers were these pioneers, wooden of course, and but sorry specimens of marine architecture, but they opened the way for great things. For some years longer the rushing torrent of the Ste. Marie's kept Lake Superior tightly closed to steamboats, but about 1840 the richness of the copper mines bordering upon that lake began to attract capital, and the need of steam navigation became crying. In 1845 men determined to put some sort of a craft upon the lake that would not be dependent upon the whims of wind and sails for propulsion. Accordingly, the sloop "Ocean," a little craft of fifteen tons, was fitted out with an engine and wheels at Detroit and towed to the "Soo." There she was dragged out of the water and made the passage between the two lakes on rollers. The "Independence," a boat of about the same size, was treated in the same way later in the year. Scarcely anything in the history of navigation, unless it be the first successful application of steam to the propulsion of boats is of equal importance with the first appearance of steamboats in Lake Superior. It may be worth while to abandon for a moment the orderly historical sequence of this narrative, to emphasize the wonderful contrast between the commerce of Lake Superior in the days of the "Independence" and now—periods separated by scarcely sixty years. To-day the commerce of that lake is more than half of all the great lakes combined. It is conducted in steel vessels, ranging from 1500 to 8500 tons, and every year sees an increase in their size. In 1901 more than 27,000,000 tons of freight were carried in Lake Superior vessels, a gain of nearly 3,000,000 over the year before. The locks in the "Soo" canal, of which more later, have twice had to be enlarged, while the Canadian Government has built a canal of its own on the other side of the river. The discovery and development of the wonderful deposits of iron ore at the head of the lake have proved the greatest factors in the upbuilding of its commerce, and the necessity for getting this ore to the mills in Illinois, Ohio, and Pennsylvania, has resulted in the creation of a class of colossal cargo-carriers on the lake that for efficiency and results, though not for beauty, outdo any vessel known to maritime circles.

A VANISHING TYPE ON THE LAKES

At the present time, when the project of a canal to connect the Atlantic and Pacific oceans at the Central American Isthmus has almost passed out of the sphere of discussion and into that of action, there is suggestiveness in the part that the canal at the "Soo" played in stimulating lake commerce. Until it was dug, the lake fleets grew but slowly, and the steamers were but few and far between. Freight rates were high, and the schooners and sloops made but slow passages. From an old bill, of about 1835, we learn that freight rates between Detroit and Cleveland, or Lake Erie points and Buffalo, were about as follows: Flour, thirty cents a barrel; all grain, ten cents a bushel; beef, pork, ashes, and whisky, thirteen cents a hundred pounds; skins and furs, thirty-one cents a hundred weight; staves, from Detroit to Buffalo, $6.25 a thousand. In 1831 there were but 111 vessels of all sorts on the lakes. In five years, the fleet had grown to 262, and in 1845, the year when the first steamer entered Lake Superior, to 493. In 1855, the year the "Soo" canal was opened, there were in commission 1196 vessels, steam and sail, on the unsalted seas. Then began the era of prodigious development, due chiefly to that canal which Henry Clay, great apostle as he was of internal improvements, said would be beyond the remotest range of settlements in the United States or in the moon.

At the head of Lake Superior are almost illimitable beds of iron ore which looks like rich red earth, and is scooped up by the carload with steam shovels. Tens of thousands of men are employed in digging this ore and transporting it to the nearest lake port—Duluth and West Superior being the largest shipping points. Railroads built and equipped for the single purpose of carrying the ore are crowded with rumbling cars day and night, and at the wharves during the eight or nine months of the year when navigation is open lie great steel ships, five hundred feet long, with a capacity of from six thousand to nine thousand tons of ore. Perhaps in no branch of marine architecture has the type best fitted to the need been so scientifically determined as in planning these ore boats. They are cargo carriers only, and all considerations of grace or beauty are rigidly eliminated from their design. The bows are high to meet and part the heavy billows of the tempestuous lakes, for they are run as late into the stormy fall and early winter season as the ice will permit. From the forward quarter the bulwarks are cut away, the high bow sheltering the forecastle with the crews, while back of it rises a deck-house of steel, containing the officers' rooms, and bearing aloft the bridge and wheel-house. Three hundred feet further aft rises another steel deck-house, above the engine, and between extends the long, flat deck, broken only by hatches every few feet, battened down almost level with the deck floor. During the summer, all too short for the work the busy iron carriers have to do, these boats are run at the top of their speed, and on schedules that make the economy of each minute essential. So they are built in such fashion as to make loading as easy and as rapid as possible. Sometimes there are as many as fourteen or sixteen hatches in one of these great ships, into each of which while loading the ore chutes will be pouring their red flood, and out of each of which the automatic unloaders at Cleveland or Erie will take ten-ton bites of the cargo, until six or seven thousand tons of iron ore may be unloaded in eight hours. The hold is all one great store-room, no deck above the vessel's floor except the main deck. No water-tight compartments or bulkheads divide it as in ocean ships, and all the machinery is placed far in the stern. The vessel is simply a great steel packing-box, with rounded ends, made strong to resist the shock of waves and the impact of thousands of tons of iron poured in from a bin as high above the floor as the roof of a three-story building. With vessels such as these, the cost of carrying ore has been reduced below the level of freight charges in any part of the world.

Yet comfort and speed are by no means overlooked. The quarters of the officers and men are superior to those provided on most of the ocean liners, and vastly better than anything offered by the "ocean tramps." Many of the ships have special guest-cabins fitted up for their owners, rivalling the cabins *de luxe* of the ocean greyhounds. The speed of the newer ships will average from fourteen to sixteen knots, and one of them in a season will make as many as twenty round trips between Duluth and Cleveland. Often one will tow two great steel barges almost as large as herself, great ore tanks without machinery of any kind and mounting two slender masts chiefly for signaling purposes, but also for use in case of being cut adrift. For a time, the use of these barges, with their great stowage capacity in proportion to their total displacement, was thought to offer the cheapest way of carrying ore. One mining company went very heavily into building these craft, figuring that every steamer could tow two or three of them, giving thus for each engine and crew a load of perhaps twenty-four thousand tons. But, seemingly, this expectation has been disappointed, for while the barges already constructed are in active use, most of the companies have discontinued building them. Indeed, at the moment of the preparation of this book, there were but two steel barges building in all the shipyards of the great lakes.

Another form of lake vessel of which great things were expected, but which disappointed its promoters, is the "whaleback," commonly called by the sailors "pigs." These are cigar-shaped craft, built of steel, their decks, from the bridge aft to the engine-house, rounded like the back of a whale, and carried only a few feet above the water. In a sea, the greater part of the deck is all awash, and a trip from the bridge to the engine-house means not only repeated duckings, but a fair chance of being swept overboard. The first of these boats, called the "101," was built in sections, the plates being forged at Cleveland, and the bow and stern built at Wilmington, Del. The completed structure was launched at Duluth. In after years she was taken to the ocean, went round Cape Horn, and was finally wrecked on the north Pacific coast. At the time of the Columbian Exposition, a large passenger-carrying whaleback, the "Christopher Columbus," was built, which still plies on Lake Michigan, though there is nothing discernible in the way of practical advantage in this design for passenger vessels. For cargo carrying there would seem to be much in the claims of their inventor, Alexander McDougall, for their superior capacity and stability, yet they have not been generally adopted. The largest whaleback now on the lakes is named after Mr. McDougall, is four hundred and thirty feet over all, fifty feet beam, and of eight thousand tons capacity. She differs from the older models in having a straight stem instead of the "pig's nose."

THE "WHALEBACK"

The iron traffic which has grown to such monster proportions, and created so noble a fleet of ships, began in 1856, when the steamer "Ontonagon" shipped two hundred and ninety-six tons of ore at Duluth. To-day, one ship of a fleet numbering hundreds will carry nine thousand tons, and make twenty trips a season. Mr. Waldon Fawcett, who has published in the "Century Magazine" a careful study of this industry, estimates the total ore cargoes for a year at about 20,000,000 tons. The ships of the ore fleet will range from three hundred and fifty to five hundred feet in length, with a draft of about eighteen feet—at which figure it must stop until harbors and channels are deepened. Their cost will average $350,000. The cargoes are worth upward of $100,000,000 annually, and the cost of transportation has been so reduced that in some instances a ton is carried twenty miles for one cent. The seamen, both on quarterdeck and forecastle, will bear comparison with their salt-water brethren for all qualities of manhood. Indeed, the lot of the sailor on the lakes naturally tends more to the development of his better qualities than does that of the salt-water jack, for he is engaged by the month, or season, rather than by the trip; he is never in danger of being turned adrift in a foreign port, nor of being "shanghaied" in a home one. He has at least three months in winter to fit himself for shore work if he desires to leave the water, and during the season he is reasonably sure of seeing his family every fortnight. A strong trades-union among the lake seamen keeps wages up and regulates conditions of employment. At the best, however, seafaring on either lake or ocean is but an ill-paid calling, and the earnings of the men who command and man the great ore-carriers are sorely out of proportion to the profits of the employing corporations. Mr. Fawcett asserts that $11,250 net earnings for a single trip was not unusual in one season, and that this sum might have been increased by $4500 had the owners taken a return cargo of coal instead of rushing back light for more ore. As the vessels of the ore fleet are owned in the main by the steel trust, their earnings are a consideration second to their efficiency in keeping the mills supplied with ore.

The great canal at Sault Ste. Marie which has caused this prodigious development of the lake shipping has been under constant construction and reconstruction for almost half a century. It had its origin in a gift of 750,000 acres of public lands from the United States Government to the State of Michigan. The State, in its turn, passed the lands on to a private company which built the canal. This work was wholly unsatisfactory, and very wisely the Government took the control of this artificial waterway out of private hands and assumed its management itself. At once it expended about $8,000,000 upon the enlargement and improvement of the canal. Scarcely was it opened before the ratio at which the traffic increased showed that it would not long be sufficient. Enlarged in 1881, it gave a capacity of from fourteen feet, nine inches to fifteen feet in depth, and with locks only four hundred feet in length. Even a ditch of this size proved of inestimable value in helping vessels to avoid the eighteen feet drop between Lake Superior and Lake Huron. By 1886 the tonnage which passed through the canal each year exceeded 9,000,000, and then for the first time this great waterway with a season limited to eight or nine months, exceeded in the volume of its traffic the great Suez canal. But shippers at once began to complain of its dimensions. Vessels were constantly increasing both in length and in draught, and the development of the great iron fields gave assurance that a new and prodigious industry would add largely to the size of the fleet, which up to that time had mainly been employed in carrying grain. Accordingly the Government rebuilt the locks until they now are one hundred feet in width, twenty-one feet deep, and twelve hundred feet long. Immediately vessels were built of a size which tests even this great capacity, and while the traffic through De Lessep's famous canal at Suez has for a decade remained almost stationary, being 9,308,152 tons, in 1900, the traffic through the "Soo" has increased in almost arithmetical proportion every year, attaining in 1901, 24,696,736 tons, or more than the combined tonnage of the Suez, Kiel, and Manchester canals, though the "Soo" is closed four months in the year. In 1887 the value of the iron ore shipments through the canal was $8,744,995. Ten years later it exceeded $30,000,000. Meanwhile it must be remembered that the Canadian Government has built on its own side of the river very commodious canals which themselves carry no small share of the Lake Superior shipments. An illustration of the fashion in which superior facilities at one end of a great line of travel compel improvements all along the line is afforded by the fact that since the canal at the "Soo" has been deepened so as to take vessels of twenty-one feet draught with practically no limit upon their length, the cry has gone up among shippers and vessel men for a twenty-foot channel from Duluth to the sea. At present there are several points in the lower lakes, notably at what is called the Lime Kiln Crossing, below Detroit, where twenty-foot craft are put to some hazard, while beyond Buffalo the shallow Welland Canal, with its short locks, and the shallow canals of the St. Lawrence River have practically stopped all effort to establish direct and profitable communication between the great lakes and the ocean. Such efforts have been made and the expedients adopted to get around natural obstacles have sometimes been almost pathetic in the story they tell of the eagerness of the lake marine to find an outlet to salt-water. Ships are cut in two at Cleveland or at Erie and sent, thus disjointed, through the canals to be patched together again at Quebec or Montreal. One body of Chicago capitalists built four steel steamers of about 2500 tons capacity each, and of dimensions suited to the locks in the Welland Canal, in the hopes of maintaining a regular freight line between that city and Liverpool. The vessels were loaded with full cargo as far as Buffalo, there discharged half their freight, and went on thus half-laden through the Canadian canals. But the loss in time and space, and the expense of reshipment of cargo made the experiment an unprofitable one. Scarcely a year has passed that some such effort has not been made, and constantly the wonderful development of the ship-building business on the Great Lakes greatly increases the vigor of the demand for an outlet. Steel ships can be built on the lakes at a materially smaller cost than anywhere along the seaboard. In the report of the Commissioner of Navigation for 1901 it is noted that more than double the tonnage of steel construction on the Atlantic coast was reported from the lakes. If lake builders could send their vessels easily and safely to the ocean, we should not need subsidies and special legislation to reestablish the American flag abroad. By the report already quoted, it is shown that thirty-nine steel steamers were built in lake yards of a tonnage ranging from 1089 tons to 5125. Wooden ship-building is practically dead on the lakes. In June of that year twenty-six more steel steamers, with an aggregate tonnage of 81,000 were on the stocks in the lake yards. Two of these are being built for ocean service, but both will have to be cut in two before they can get through the Canadian canals. It is not surprising that there appears among the people living in the commonwealths which border on the Great Lakes a certain doubt as to whether the expenditure by the United States Government of $200,000,000 for a canal at the Isthmus will afford so great a measure of encouragement to American shipping and be of as immediate advantage to the American exporter, as a twenty-foot channel from Duluth to tide-water.

Though the old salt may sneer at the freshwater sailor who scarcely need know how to box the compass, to whom the art of navigation is in the main the simple practise of steering from port to port guided by headlands and lights, who is seldom long out of sight of land, and never far from aid, yet the perils of the lakes are quite as real as those which confront the ocean seaman, and the skill and courage necessary for withstanding them quite as great as his. The sailor's greatest safeguard in time of tempest is plenty of searoom. This the lake navigator never has. For him there is always the dreaded lee shore only a few miles away. Anchorage on the sandy bottom of the lakes is treacherous, and harbors are but few and most difficult of access. Where the ocean sailor finds a great bay, perhaps miles in extent, entered by a gateway thousands of yards across, offering a harbor of refuge in time of storm, the lake navigator has to run into the narrow mouth of a river, or round under the lee of a government breakwater hidden from sight under the crested waves and offering but a precarious shelter at best. Chicago, Cleveland, Milwaukee—most of the lake ports have witnessed such scenes of shipwreck and death right at the doorway of the harbor, as no ocean port could tell. At Chicago great schooners have been cast far up upon the boulevard that skirts a waterside park, or thrown bodily athwart the railroad tracks that on the south side of the city border the lake. The writer has seen from a city street, crowded with shoppers on a bright but windy day, vessels break to pieces on the breakwater, half a mile away but in plain sight, and men go down to their death in the raging seas. On all the lakes, but particularly on the smaller ones, an ugly sea is tossed up by the wind in a time so short as to seem miraculous to the practised navigator of the ocean. The shallow water curls into breakers under the force of even a moderate wind, and the vessels are put to such a strain, in their struggles, as perhaps only the craft built especially for the English channel have to undergo. Some of the most fatal disasters the lakes have known resulted from iron vessels, thus racked and tossed, sawing off, as the phrase goes, the rivets that bound their plates together, and foundering. Fire, too, has numbered its scores of victims on lake steamers, though this danger, like indeed most others, is greatly decreased by the increased use of steel as a structural material and the great improvement in the model of the lake craft. Even ten years ago the lake boats were ridiculous in their clumsiness, their sluggishness, and their lack of any of the charm and comfort that attend ocean-going vessels, but progress toward higher types has been rapid, and there are ships on the lakes to-day that equal any of their size afloat.

For forty years it has been possible to say annually, "This is the greatest year in the history of the lake marine." For essentially it is a new and a growing factor in the industrial development of the United States. So far, from having been killed by the prodigious development of our railroad system, it has kept pace with that system, and the years that have seen the greatest number of miles of railroad built, have witnessed the launching of the biggest lake vessels. There is every reason to believe that this growth will for a long time be persistent, that the climax has not yet been reached. For it is incredible that the Government will permit the barrier at Niagara to the commerce of these great inland seas to remain long unbroken. Either by the Mohawk valley route, now followed by the Erie canal, or by the route down the St. Lawrence, with a deepening and widening of the present Canadian canals, and a new canal down from the St. Lawrence to Lake Champlain, a waterway will yet be provided. The richest coast in the world is that bordering on the lakes. The cheapest ships in the world can there be built. Already the Government has spent its tens and scores of millions in providing waterways from the extreme northwest end to the southeastern extremity of this water system, and it is unbelievable that it shall long remain violently stopped there. New devices for digging canals; such as those employed in the Chicago drainage channel, and the new pneumatic lock, the power and capacity of which seem to be practically unlimited, have vastly decreased the cost of canal building, and multiplied amazingly the value of artificial waterways. As it is admitted that the greatness and the wealth of New York State are much to be credited to the Erie canal, so the prosperity and populousness of the whole lake region will be enhanced when lake sailors and the lake ship-builders are given a free waterway to the ocean.

CHAPTER VIII

THE MISSISSIPPI AND TRIBUTARY RIVERS—THE CHANGING PHASES OF THEIR SHIPPING—RIVER NAVIGATION AS A NATION-BUILDING FORCE—THE VALUE OF SMALL STREAMS—WORK OF THE OHIO COMPANY—AN EARLY PROPELLER—THE FRENCH FIRST ON THE MISSISSIPPI—THE SPANIARDS AT NEW ORLEANS—EARLY METHODS OF NAVIGATION—THE FLATBOAT, THE BROADHORN, AND THE KEELBOAT—LIFE OF THE RIVERMEN—PIRATES AND BUCCANEERS—LAFITTE AND THE BARATARIANS—THE GENESIS OF THE STEAMBOATS—CAPRICIOUS RIVER—FLUSH TIMES IN NEW ORLEANS—RAPID MULTIPLICATION OF STEAMBOATS—RECENT FIGURES ON RIVER SHIPPING—COMMODORE WHIPPLE'S EXPLOIT—THE MEN WHO STEERED THE STEAMBOATS—THEIR TECHNICAL EDUCATION—THE SHIPS THEY STEERED—FIRES AND EXPLOSIONS—HEROISM OF THE PILOTS—THE RACERS.

It is the ordinary opinion, and one expressed too often in publications which might be expected to speak with some degree of accuracy, that river transportation in the United States is a dying industry. We read every now and then of the disappearance of the magnificent Mississippi River steamers, and the magazines not infrequently treat their readers to glowing stories of what is called the "flush" times on the Mississippi, when the gorgeousness of the passenger accommodations, the lavishness of the table, the prodigality of the gambling, and the mingled magnificence and outlawry of life on the great packets made up a picturesque and romantic phase of American life. It is true that much of the picturesqueness and the romance has departed long since. The great river no longer bears on its turbid bosom many of the towering castellated boats built to run, as the saying was, on a heavy dew, but still carrying their tiers upon tiers of ivory-white cabins high in air. The time is past when the river was the great passenger thoroughfare from St. Louis to New Orleans. Some few packets still ply upon its surface, but in the main the passenger traffic has been diverted to the railroads which closely parallel its channel on either side. The American travels much, but he likes to travel fast, and for passenger traffic, except on a few routes where special conditions obtain, the steamboat has long since been outclassed by the railroads.

Yet despite the disappearance of its spectacular conditions the water traffic on the rivers of the Mississippi Valley is greater now than at any time in its history. Its methods only have changed. Instead of gorgeous packets crowded with a gay and prodigal throng of travelers for pleasure, we now find most often one dingy, puffing steamboat, probably with no passenger accommodations at all, but which pushes before her from Pittsburg to New Orleans more than a score of flatbottomed, square-nosed scows, aggregating perhaps more than an acre of surface, and heavy laden with coal. Such a tow—for "tow" it is in the river vernacular, although it is pushed—will transport more in one trip than would suffice to load six heavy freight trains. Not infrequently the barges or scows will number more than thirty, carrying more than 1000 tons each, or a cargo exceeding in value $100,000. During the season when navigation is open on the Ohio and its tributaries, this traffic is pursued without interruption. Through it and through the local business on the lower Mississippi, and the streams which flow into it, there is built up a tonnage which shows the freight movement, at least, on the great rivers, to exceed, even in these days of railroads, anything recorded in their history.

No physical characteristic of the United States has contributed so greatly to the nationalization of the country and its people, as the topography of its rivers. From the very earliest days they have been the pathways along which proceeded exploration and settlement. Our forefathers, when they found the narrow strip of land along the Atlantic coast which they had at first occupied, becoming crowded, according to their ideas at the time, began working westward, following the river gaps. Up the Hudson and westward by the Mohawk, up the Susquehanna and the Potomac, carrying around the falls that impeded the course of those streams, trudging over the mountains, and building flatboats at the headwaters of the Ohio, they made their way west. Some of the most puny streams were utilized for water-carriers, and the traveler of to-day on certain of the railroads through western New York and Pennsylvania, will be amazed to see the remnants of canals, painfully built in the beds of brawling streams, that now would hardly float an Indian birch-bark canoe. In their time these canals served useful purposes. The stream was dammed and locked every few hundred yards, and so converted into a placid waterway with a flight of mechanical steps, by which the boats were let down to, or raised up from tidewater. To-day nothing remains of most of these works of engineering, except masses of shattered masonry. For the railroads, using the river's bank, and sometimes even part of the retaining walls of the canals for their roadbeds, have shrewdly obtained and swiftly employed authority to destroy all the fittings of these waterways which might, perhaps, at some time, offer to their business a certain rivalry.

The corporation known as the Ohio Company, with a great purchase of land from Congress in 1787, by keen advertising, and the methods of the modern real-estate boomer, started the tide of emigration and the fleet of boats down the Ohio. The first craft sent out by this corporation was named, appropriately enough, the "Mayflower." She drifted from Pittsburg to a spot near the mouth of the Muskingum river. Soon the immigrants began to follow by scores, and then by thousands. Mr. McMaster has collected some contemporary evidence of their numbers. One man at Fort Pitt saw fifty flatboats set forth between the first of March and the middle of April, 1787.

Between October, 1786, and May, 1787—the frozen season when boats were necessarily infrequent—the adjutant at Fort Harmer counted one hundred and seventy-seven flat-boats, and estimated they carried twenty-seven hundred settlers. A shabby and clumsy fleet it was, indeed, with only enough seamanship involved to push off a sand-bar, but it was a great factor in the upbuilding of the nation. And a curious fact is that the voyagers on one of these river craft hit upon the principle of the screw-propeller, and put it to effective use. The story is told in the diary of Manasseh Cutler, a member of the Ohio Company, who writes: "Assisted by a number of people, we went to work and constructed a machine in the form of a screw, with short blades, and placed it in the stern of the boat, which we turned with a crank. It succeeded to perfection, and I think it a very useful discovery." But the discovery was forgotten for nearly three-quarters of a century, until John Ericsson rediscovered and utilized it.

Once across the divide, the early stream of immigration took its way down the Ohio River to the Mississippi. There it met the outposts of French power, for the French burst open that great river, following their missionaries, Marquette and Joliet, down from its headwaters in Wisconsin, or pressing up from their early settlements at New Orleans. Doubtless, if it had not been that the Mississippi afforded the most practicable, and the most useful highway from north to south, the young American people would have had a French State to the westward of them until they had gone much further on the path toward national manhood. But the navigation of the Mississippi and its tributaries was so rich a prize, that it stimulated alike considerations of individual self-interest and national ambition. From the day when the first flatboat made its way from the falls of the Ohio to New Orleans, it was the fixed determination of all people living by the great river, or using it as a highway for commerce, that from its headwaters to its mouth it should be a purely American stream. It was in this way that the Mississippi and its tributaries proved to be, as I have said, a great influence in developing the spirit of coherent nationality among the people of the young nation.

Indeed, no national Government could be of much value to the farmers and trappers of Kentucky and Tennessee that did not assure them the right to navigate the Mississippi to its mouth, and find there a place to trans-ship their goods into ocean-going vessels. From the Atlantic seaboard they were shut off by a wall, that for all purpose of export trade was impenetrable. The swift current of the rivers beat back their vessels, the towering ranges of the Alleghanies mocked at their efforts at road building. From their hills flowed the water that filled the Father of Waters and his tributaries. Nature had clearly designed this for their outlet. As James Madison wrote: "The Mississippi is to them everything. It is the Hudson, the Delaware, the Potomac, and all the navigable waters of the Atlantic coast formed into one stream." Yet, when the first trader, in 1786, drifted with his flatboat from Ohio down to New Orleans, thus entering the confines of Spanish territory, he was seized and imprisoned, his goods were taken from him, and at last he was turned loose, penniless, to plod on foot the long way back to his home, telling the story of his hardships as he went along. The name of that man was Thomas Amis, and after his case became known in the great valley, it ceased to be a matter of doubt that the Americans would control the Mississippi. He was in a sense the forerunner of Jefferson and Jackson, for after his time no intelligent statesman could doubt that New Orleans must be ours, nor any soldier question the need for defending it desperately against any foreign power. The story of the way in which Gen. James Wilkinson, by intrigue and trickery, some years later secured a partial relaxation of Spanish vigilance, can not be told here, though his plot had much to do with opening the great river.

FLATBOATS MANNED WITH RIFLEMEN

The story of navigation on the Mississippi River, is not without its elements of romance, though it does not approach in world interest the story of the achievements of the New England mariners on all the oceans of the globe. Little danger from tempest was encountered here. The natural perils to navigation were but an ignoble and unromantic kind—the shifting sand-bar and the treacherous snag. Yet, in the early days, when the flatboats were built at Cincinnati or Pittsburg, with high parapets of logs or heavy timber about their sides, and manned not only with men to work the sweeps and hold the steering oar, but with riflemen, alert of eye, and unerring of aim, to watch for the lurking savage on the banks, there was peril in the voyage that might even affect the stout nerves of the hardy navigator from New Bedford or Nantucket. For many long years in the early days of our country's history, the savages of the Mississippi Valley were always hostile, continually enraged.

The French and the English, bent upon stirring up antagonism to the growing young nation, had their agents persistently at work awakening Indian hostility, and, indeed, it is probable that had this not been the case, the rough and lawless character of the American pioneers, and their entire indifference to the rights of the Indians, whom they were bent on displacing, would have furnished sufficient cause for conflict.

First of the craft to follow the Indian canoes and the bateaux of the French missionaries down the great rivers, was the flatboat—a homely and ungraceful vessel, but yet one to which the people of the United States owe, perhaps, more of real service in the direction of building up a great nation than they do to Dewey's "Olympia," or Schley's "Brooklyn." A typical flatboat of the early days of river navigation was about fifty-five feet long by sixteen broad. It was without a keel, as its name would indicate, and drew about three feet of water. Amidships was built a rough deck-house or cabin, from the roof of which extended on either side, two long oars, used for directing the course of the craft rather than for propulsion, since her way was ever downward with the current, and dependent upon it. These great oars seemed to the fancy of the early flatboat men, to resemble horns, hence the name "broadhorns," sometimes applied to the boats. Such a boat the settler would fill with household goods and farm stock, and commit himself to the current at Pittsburg. From the roof of the cabin that housed his family, cocks crew and hens cackled, while the stolid eyes of cattle peered over the high parapet of logs built about the edge for protection against the arrow or bullet of the wandering redskin. Sometimes several families would combine to build one ark. Drifting slowly down the river—the voyage from Pittsburg to the falls of the Ohio, where Louisville now stands, requiring with the best luck, a week or ten days—the shore on either hand would be closely scanned for signs of unusual fertility, or for the opening of some small stream suggesting a good place to "settle." When a spot was picked out the boat would be run aground, the boards of the cabin erected skilfully into a hut, and a new outpost of civilization would be established. As these settlements multiplied, and the course of emigration to the west and southwest increased, river life became full of variety and gaiety. In some years more than a thousand boats were counted passing Marietta. Several boats would lash together and make the voyage to New Orleans, which sometimes occupied months, in company. There would be frolics and dances, the notes of the violin—an almost universal instrument among the flatboat men—sounded across the waters by night to the lonely cabins on the shores, and the settlers not infrequently would put off in their skiffs to meet the unknown voyagers, ask for the news from the east, and share in their revels. Floating shops were established on the Ohio and its tributaries—flatboats, with great cabins fitted with shelves and stocked with cloth, ammunition, tools, agricultural implements, and the ever-present whisky, which formed a principal staple of trade along the rivers. Approaching a clump of houses on the bank, the amphibious shopkeeper would blow lustily upon a horn, and thereupon all the inhabitants would flock down to the banks to bargain for the goods that attracted them. As the population increased the floating saloon and the floating gambling house were added to the civilized advantages the river bore on its bosom. Trade was long a mere matter of barter, for currency was seldom seen in these outlying settlements. Skins and agricultural products were all the purchasers had to give, and the merchant starting from Pittsburg with a cargo of manufactured goods, would arrive at New Orleans, perhaps three months later, with a cabin filled with furs and a deck piled high with the products of the farm. Here he would dispose of his cargo, perhaps for shipment to Europe, sell his flatboat for the lumber in it, and begin his painful way back again to the head of navigation.

The flatboat never attempted to return against the stream. For this purpose keel-boats or barges were used, great hulks about the size of a small schooner, and requiring twenty-five men at the poles to push one painfully up stream. Three methods of propulsion were employed. The "shoulder pole," which rested on the bottom, and which the boatman pushed, walking from bow to stern as he did so; tow-lines, called cordelles, and finally the boat was drawn along by pulling on overhanging branches. The last method was called "bushwhacking." These became in time the regular packets of the rivers, since they were not broken up at the end of the voyage and required trained crews for their navigation. The bargemen were at once the envy and terror of the simple folk along the shores. A wild, turbulent class, ready to fight and to dance, equally enraptured with the rough scraping of a fiddle by one of their number, or the sound of the war-whoop, which promised the only less joyous diversion of a fight, they aroused all the inborn vagrant tendencies of the riverside boys, and to run away with a flatboat became, for the Ohio or Indiana lad, as much of an ambition as to run away to sea was for the boy of New England. It will be remembered that Abraham Lincoln for a time followed the calling of a flatboatman, and made a voyage to New Orleans, on which he first saw slaves, and later invented a device for lifting flatboats over sand-bars, the model for which is still preserved at Washington, though the industry it was designed to aid is dead. Pigs, flour, and bacon, planks and shingles, ploughs, hoes, and spades, cider and whisky, were among the simple articles dealt in by the owners of the barges. Their biggest market was New Orleans, and thither most of their food staples were carried, but for agricultural implements and whisky there was a ready sale all along the route. Tying up to trade, or to avoid the danger of night navigation, the boatmen became the heroes of the neighborhood. Often they invited all hands down to their boat for a dance, and by flaring torches to the notes of accordion and fiddle, the evening would pass in rude and harmless jollity, unless too many tin cups or gourds of fiery liquor excited the always ready pugnacity of the men. They were ready to brag of their valor, and to put their boasts to the test. They were "half horse, half alligator," according to their own favorite expression, equally prepared with knife or pistol, fist, or the trained thumb that gouged out an antagonist's eye, unless he speedily called for mercy. "I'm a Salt River roarer!" bawled one in the presence of a foreign diarist. "I can outrun, outjump, throw down, drag out and lick any man on the river! I love wimmen, and I'm chock full of fight!" In every crew the "best" man was entitled to wear a feather or other badge, and the word "best" had no reference to moral worth, but merely expressed his demonstrated ability to whip any of his shipmates. They had their songs, too, usually sentimental, as the songs of rough men are, that they bawled out as they toiled at the sweeps or the pushpoles. Some have been preserved in history:

> "It's oh! As I was walking out,
>
> One morning in July,
>
> I met a maid who axed my trade.
>
> 'A flatboatman,' says I.
>
> "And it's oh! She was so neat a maid
>
> That her stockings and her shoes
>
> She toted in her lily-white hands,
>
> For to keep them from the dews."

"THE EVENING WOULD PASS IN RUDE AND HARMLESS JOLLITY."

Just below the mouth of the Wabash on the Ohio was the site of Shawneetown, which marked the line of division between the Ohio and the

Mississippi trade. Here goods and passengers were debarked for Illinois, and here the Ohio boatmen stopped before beginning their return trip. Because of the revels of the boatmen, who were paid off there, the place acquired a reputation akin to that which Port Said, at the northern entrance to the Suez Canal, now holds. It held a high place in river song and story.

> "Some row up, but we row down,
>
> All the way to Shawneetown.
>
> Pull away, pull away,"

was a favorite chorus.

Natchez, Tennessee, held a like unsavory reputation among the Mississippi River boatmen, for there was the great market in which were exchanged northern products for the cotton, yams, and sugar of the rich lands of the South.

For food on the long voyage, the boatmen relied mostly on their rifles, but somewhat on the fish that might be brought up from the depths of the turbid stream, and the poultry and mutton which they could secure from the settlers by barter, or not infrequently, by theft. Wild geese were occasionally shot from the decks, while a few hours' hunt on shore would almost certainly bring reward in the shape of wild turkey or deer. A somewhat archaic story among river boatmen tells of the way in which "Mike Fink," a famous character among them, secured a supply of mutton. Seeing a flock of sheep grazing near the shore, he ran his boat near them, and rubbed the noses of several with Scotch snuff. When the poor brutes began to caper and sneeze in dire discomfort, the owner arrived on the scene, and asked anxiously what could ail them. The bargeman, as a traveled person, was guide, philosopher, and friend to all along the river, and so, when informed that his sheep were suffering from black murrain, and that all would be infected unless those already afflicted were killed, the farmer unquestioningly shot those that showed the strange symptoms, and threw the bodies into the river, whence they were presently collected by the astute "Mike," and turned into fair mutton for himself and passengers. Such exploits as these added mightily to the repute of the rivermen for shrewdness, and the farmer who suffered received scant sympathy from his neighbors.

But the boatmen themselves had dangers to meet, and robbers to evade or to outwit. At any time the lurking Indian on the banks might send a death-dealing arrow or bullet from some thicket, for pure love of slaughter. For a time it was a favorite ruse of hostiles, who had secured a white captive, to send him alone to the river's edge, under threat of torture, there to plead with outstretched hands for aid from the passing raft. But woe to the mariner who was moved by the appeal, for back of the unfortunate, hidden in the bushes, lay ambushed savages, ready to leap upon any who came ashore on the errand of mercy, and in the end neither victim nor decoy escaped the fullest infliction of redskin barbarity. There were white outlaws along the rivers, too; land pirates ready to rob and murder when opportunity offered, and as the Spanish territory about New Orleans was entered, the dangers multiplied. The advertisement of a line of packets sets forth:

"No danger need be apprehended from the enemy, as every person whatever will be under cover, made proof against rifle or musket balls, and convenient portholes for firing out of. Each of the boats are armed with six pieces, carrying a pound ball, also a number of muskets, and amply supplied with ammunition, strongly manned with choice hands, and masters of approved knowledge."

The English of the advertisement is not of the most luminous character, yet it suffices to tell clearly enough to any one of imagination, the story of some of the dangers that beset those who drifted from Ohio to New Orleans.

The lower reaches of the Mississippi River bore among rivermen, during the early days of the century, very much such a reputation as the Spanish Main bore among the peaceful mariners of the Atlantic trade. They were the haunts of pirates and buccaneers, mostly ordinary cheap freebooters, operating from the shore with a few skiffs, or a lugger, perhaps, who would dash out upon a passing vessel, loot it, and turn it adrift. But one gang of these river pirates so grew in power and audacity, and its leaders so ramified their associations and their business relations, as for a time to become a really influential factor in the government of New Orleans, while for a term of years they even put the authority of the United States at nought. The story of the brothers Lafitte and their nest of criminals at Barataria, is one of the most picturesque in American annals. On a group of those small islands crowned with live-oaks and with fronded palms, in that strange waterlogged country to the southwest of the Crescent City, where the sea, the bayou, and the marsh fade one into the other until the line of demarkation can scarcely be traced, the Lafittes established their colony. There they built cabins and storehouses, threw up-earthworks, and armed them with stolen cannon. In time the plunder of scores of vessels filled the warehouses with the goods of all nations, and as the wealth of the colony grew its numbers increased. To it were attracted the adventurous spirits of the creole city. Men of Spanish and of French descent, negroes, and quadroons, West Indians from all the islands scattered between North and South America, birds of prey, and fugitives from justice of all sorts and kinds, made that a place of refuge. They brought their women and children, and their slaves, and the place became a small principality, knowing no law save Lafitte's will. With a fleet of small schooners the pirates would sally out into the Gulf and plunder vessels of whatever sort they might encounter. The road to their hiding-place was difficult to follow, either in boats or afoot, for the tortuous bayous that led to it were intertwined in an almost inextricable maze, through which, indeed, the trained pilots of the colony picked their way with ease, but along which no untrained helmsman could follow them. If attack were made by land, the marching force was confronted by impassable rivers and swamps; if by boats, the invaders pressing up a channel which seemed to promise success, would find themselves suddenly in a blind alley, with nothing to do save to retrace their course. Meanwhile, for the greater convenience of the pirates, a system of lagoons, well known to them, and easily navigated in luggers, led to the very back door of New Orleans, the market for their plunder. Of the brothers Lafitte, one held state in the city as a successful merchant, a man not without influence with the city government, of high standing in the business community, and in thoroughly good repute. Yet he was, in fact, the agent for the pirate colony, and the goods he dealt in were those which the picturesque ruffians of Barataria had stolen from the vessels about the mouth of the Mississippi River. The situation persisted for nearly half a score of years. If there were merchants, importers and shipowners in New Orleans who suffered by it, there were others who profited by it, and it has usually been the case that a crime or an injustice by which any considerable number of people profit, becomes a sort of vested right, hard to disturb. And, indeed, the Baratarians were not without a certain rude sense of patriotism and loyalty to the United States, whose laws they persistently violated. For when the second war with Great Britain was declared and Packenham was dispatched to take New Orleans, the commander of the British fleet made overtures to Lafitte and his men, promising them a liberal subsidy and full pardon for all past offenses, if they would but act as his allies and guide the British invaders to the most vulnerable point in the defenses of the Crescent City. The offer was refused, and instead, the chief men of the pirate colony went straightway to New Orleans to put Jackson on his guard, and when the opposing forces met on the plains of Chalmette, the very center of the American line was held by Dominique Yon, with a band of his swarthy Baratarians, with howitzers which they themselves had dragged from their pirate stronghold to train upon the British. Many of us, however law-abiding, will feel a certain sense that the romance of history would have been better served, if after this act of patriotism, the pirates had been at least peacefully dispersed. But they were wedded to their predatory life, returned with renewed zeal to their piracies, and were finally destroyed by the State forces and a United States naval expedition, which burned their settlement, freed their slaves, razed their fortifications, confiscated their cannon, killed many of their people, and dispersed the rest among the swamps and forests of southern Louisiana.

In 1809 a New York man, by name Nicholas J. Roosevelt, set out from Pittsburg in a flatboat of the usual type, to make the voyage to New Orleans. He carried no cargo of goods for sale, nor did he convey any band of intended settlers, yet his journey was only second in importance to the ill-fated one, in which the luckless Amis proved that New Orleans must be United States territory, or the wealth of the great interior plateau would be effectively bottled up. For Roosevelt was the partner of Fulton and Livingston in their new steamboat enterprise, having himself suggested the vertical paddle-wheel, which for more than a half a century was the favorite means of utilizing steam power for the propulsion of boats. He was firm in the belief that the greatest future for the steamboat was on the great rivers that tied together the rapidly growing commonwealths of the middle west, and he undertook this voyage for the purpose of studying the channel and the current of the rivers, with the view to putting a steamer on them. Wise men assured him that on the upper river his scheme was destined to failure. Could a boat laden with a heavy engine be made of so light a draught as to pass over the shallows of the Ohio? Could it run the falls at Louisville, or be dragged around them as the flatboats often were? Clearly not. The only really serviceable type of river craft was the flatboat, for it would go where there was water enough for a muskrat to swim in, would glide unscathed over the concealed snag or, thrusting its corner into the soft mud of some protruding bank, swing around and go on as well stern first as before. The flatboat was the sum of human ingenuity applied to river navigation. Even barges were proving failures and passing into disuse, as the cost of poling them upstream was greater than any profit to be reaped from the voyage. Could a boat laden with thousands of pounds of machinery make her way northward against that swift current? And if not, could steamboat men be continually taking expensive engines down to New Orleans and abandoning them there, as the old-time river men did their rafts and scows? Clearly not. So Roosevelt's appearance on the river did not in any way disquiet the flatboatmen, though it portended their disappearance as a class. Roosevelt, however, was in no wise discouraged. Week after week he drifted along the Ohio and Mississippi, taking detailed soundings, studying the course of the current, noting the supply of fuel along the banks, observing the course of the rafts and flatboats as they drifted along at the mercy of the tide. Nothing escaped his attention, and yet it may well be doubted whether the mass of data he collected was in fact of any practical value, for the great river is the least understandable of streams. Its channel is as shifting as the mists above Niagara. Where yesterday the biggest boat on the river, deep laden with cotton, might pass with safety, there may be to-day a sand-bar scarcely hidden beneath the tide. Its banks change over night in form and in appearance. In time of flood it cuts new channels for itself, leaving in a few days river towns far in the interior, and suddenly giving a water frontage to some plantation whose owner had for years mourned over his distance from the river bank. Capricious and irresistible, working insidiously night and day, seldom showing the progress of its endeavors until some huge slice of land, acres in extent, crumbles into the flood, or some gully or cut-off all at once appears as the main channel, the Mississippi, even now when the Government is at all times on the alert to hold it in bounds, is not to be lightly learned nor long trusted. In Roosevelt's time, before the days of the river commission, it must have been still more difficult to comprehend. Nevertheless, the information he collected, satisfied him that the stream was navigable for steamers, and his report determined his partners to build the pioneer craft at Pittsburg. She was completed, "built after the fashion of a ship with portholes in her side," says a writer of the time, dubbed the "Orleans," and in 1812 reached the city on the sodden prairies near the mouth of the Mississippi, whose name we now take as a synonym for quaintness, but which at that time had seemingly the best chance to become a rival of London and Liverpool, of any American town. For just then the great possibilities of the river highway were becoming apparent. The valley was filling up with farmers, and their produce sought the shortest way to tide-water. The streets of the city were crowded with flatboatmen, from Indiana, Ohio, and Kentucky, and with sailors speaking strange tongues, and gathered from all the ports of the world. At the broad levee floated the ships of all nations. All manual work was done by the negro slaves, and already the planters were beginning to show signs of that prodigal prosperity, which, in the flush times, made New Orleans the gayest city in the United States. In 1813 Jackson put the final seal on the title-deeds to New Orleans, and made the Mississippi forever an American river by defeating the British just outside the city's walls, and then river commerce grew apace. In 1817 fifteen hundred flatboats and five hundred barges tied up to the levee. By that time the steamboat had proved her case, for the "New Orleans" had run for years between Natchez and the Louisiana city, charging a fare of eighteen dollars for the down, and twenty-five dollars for the up trip, and earning for her owners twenty thousand dollars profits in one year. She was snagged and lost in 1814, but by that time others were in the field, first of all the "Comet," a stern-wheeler of twenty-five tons, built at Pittsburg, and entering the New Orleans-Natchez trade in 1814. The "Vesuvius," and the "Ætna."—volcanic names which suggested the explosive end of too many of the early boats—were next in the field, and the latter won fame by being the first boat to make the up trip from New Orleans to Louisville. Another steamboat, the "Enterprise," carried a cargo of, powder and ball from Pittsburg to General Jackson at New Orleans, and after some service on southern waters, made the return trip to Louisville in twenty-five days. This was a great achievement, and hailed by the people of the Kentucky town as the certain forerunner of commercial greatness, for at one time there were tied to the bank the "Enterprise" from New Orleans, the "Despatch" from

Pittsburg, and the "Kentucky Elizabeth" from the upper Kentucky River. Never had the settlement seemed to be so thoroughly in the heart of the continent. Thereafter river steamboating grew so fast that by 1819 sixty-three steamers, of varying tonnage from twenty to three hundred tons, were plying on the western rivers. Four had been built at New Orleans, one each at Philadelphia, New York, and Providence, and fifty-six on the Ohio. The upper reaches of the Mississippi still lagged in the race, for most of the boats turned off up the Ohio River, into the more populous territory toward the east. It was not until August, 1817, that the "General Pike," the first steamer ever to ascend the Mississippi River above the mouth of the Ohio, reached St. Louis. No pictures, and but scant descriptions of this pioneer craft, are obtainable at the present time. From old letters it is learned that she was built on the model of a barge, with her cabin situated on the lower deck, so that its top scarcely showed above the bulwarks. She had a low-pressure engine, which at times proved inadequate to stem the current, and in such a crisis the crew got out their shoulder poles and pushed her painfully up stream, as had been the practice so many years with the barges. At night she tied up to the bank. Only one other steamer reached St. Louis in the same twelve months. By way of contrast to this picture of the early beginnings of river navigation on the upper Mississippi, we may set over some facts drawn from recent official publications concerning the volume of river traffic, of which St. Louis is now the admitted center. In 1890 11,000,000 passengers were carried in steamboats on rivers of the Mississippi system. The Ohio and its tributaries, according to the census of that year, carried over 15,000,000 tons of freight annually, mainly coal, grain, lumber, iron, and steel. The Mississippi carries about the same amount of freight, though on its turbid tide, cotton and sugar, in no small degree, take the place of grain and the products of the furnaces and mills.

But it was a long time before steam navigation approached anything like these figures, and indeed, many years passed before the flatboat and the barge saw their doom, and disappeared. In 1821, ten years after the first steamboat arrived at New Orleans, there was still recorded in the annals of the town, the arrival of four hundred and forty-one flatboats, and one hundred and seventy-four barges. But two hundred and eighty-seven steamboats also tied up to the levee that year, and the end of the flatboat days was in sight. Ninety-five of the new type of vessels were in service on the Mississippi and its tributaries, and five were at Mobile making short voyages on the Mississippi Sound and out into the Gulf. They were but poor types of vessels at best. At first the shortest voyage up the river from New Orleans to Shippingport—then a famous landing, now vanished from the map—was twenty-two days, and it took ten days to come down. Within six years the models of the boats and the power of the engines had been so greatly improved that the up trip was made in twelve days, and the down in six. Even the towns on the smaller streams tributary to the great river, had their own fleets. Sixteen vessels plied between Nashville and New Orleans. The Red River, and even the Missouri, began to echo to the puffing of the exhaust and the shriek of the steam-whistle. Indeed, it was not very long before the Missouri River became as important a pathway for the troops of emigrants making for the great western plains and in time for the gold fields of California, as the Ohio had been in the opening days of the century for the pioneers bent upon opening up the Mississippi Valley. The story of the Missouri River voyage, the landing place at Westport, now transformed into the great bustling city of Kansas City, and all the attendant incidents which led up to the contest in Kansas and Nebraska, forms one of the most interesting, and not the least important chapters in the history of our national development.

The decade during which the steamboats and the flatboats still struggled for the mastery, was the most picturesque period of Mississippi River life. Then the river towns throve most, and waxed turbulent, noisy, and big, according to the standards of the times. Places which now are mere names on the map, or have even disappeared from the map altogether, were great trans-shipping points for goods on the way to the sea. New Madrid, for example, which nowadays we remember chiefly as being one of the stubborn obstacles in the way of the Union opening of the river in the dark days of the Civil War, was in 1826 like a seaport. Flatboats in groups and fleets came drifting to its levees heavy laden with the products of the west and south, the output of the northern farms and mills, and the southern plantations. On the crowded river bank would be disembarked goods drawn from far-off New England, which had been dragged over the mountains and sent down the Ohio to the Mississippi; furs from northern Minnesota or Wisconsin; lumber in the rough, or shaped into planks, from the mills along the Ohio; whisky from Kentucky, pork and flour from Illinois, cattle, horses, hemp, fabrics, tobacco, everything that men at home or abroad, could need or crave, was gathered up by enterprising traders along three thousand miles of waterway, and brought hither by clumsy rafts and flatboats, and scarcely less clumsy steamboats, for distribution up and down other rivers, and shipment to foreign lands.

At New Orleans there was a like deposit of all the products of that rich valley, an empire in itself. There grain, cotton, lumber, live stock, furs, the output of the farms and the spoils of the chase, were transferred to ocean-going ships and sent to foreign markets. Speculative spirits planned for the day, when this rehandling of cargoes at the Crescent City would be no longer necessary, but ships would clear from Louisville or St. Louis to Liverpool or Hamburg direct. A fine type of the American sailor, Commodore Whipple, who had won his title by good sea-fighting in the Revolutionary War, gave great encouragement to this hope, in 1800, by taking the full-rigged ship "St.

Clair," with a cargo of pork and flour, from Marietta, Ohio, down the Ohio, over the falls at Louisville, thence down the Mississippi, and round by sea to Havana, and so on to Philadelphia. This really notable exploit—to the success of which good luck contributed almost as much as good seamanship—aroused the greatest enthusiasm. The Commodore returned home overland, from Philadelphia. His progress, slow enough, at best, was checked by ovations, complimentary addresses, and extemporized banquets. He was *the* man of the moment. The poetasters, who were quite as numerous in the early days of the republic, as the true poets were scarce, signalized his exploit in verse.

"The Triton crieth,

'Who cometh now from shore?'

Neptune replieth,

"Tis the old Commodore.

Long has it been since I saw him before.

In the year '75 from Columbia he came,

The pride of the Briton, on ocean to tame.

"'But now he comes from western woods,

Descending slow, with gentle floods,

The pioneer of a mighty train,

Which commerce brings to my domain.'"

But Neptune and the Triton had no further occasion to exchange notes of astonishment upon the appearance of river-built ships on the ocean. The "St. Clair" was the first and last experiment of the sort. Late in the nineties, the United States Government tried building a torpedo-boat at Dubuque for ocean service, but the result was not encouraging.

Year after year the steamboats multiplied, not only on the rivers of the West, but on those leading from the Atlantic seaboard into the interior. It may be said justly that the application of steam to purposes of navigation made the American people face fairly about. Long they had stood, looking outward, gazing across the sea to Europe, their sole market, both for buying and for selling. But now the rich lands beyond the mountains, inviting settlers, and cut up by streams which offered paths for the most rapid and comfortable method of transportation then known, commanded their attention.

Immigrants no longer stopped in stony New England, or in Virginia, already dominated by an aristocratic land-owning class, but pressed on to Kentucky, Ohio, Tennessee, and Illinois. As the lands filled up, the little steamers pushed their noses up new streams, seeking new markets. The Cumberland, and the Tennessee, the Missouri, the Arkansas, the Red, the Tombigbee, and the Chattahoochee were stirred by the churning wheels, and over-their forests floated the mournful sough of the high-pressure exhaust.

In 1840, a count kept at Cairo, showed 4566 vessels had passed that point during the year. By 1848, a "banner" year, in the history of navigation on the Mississippi, traffic was recorded thus:

25 vessels plying between Louisville, New Orleans and Cincinnati	8,484 tons
7 between Nashville and New Orleans	2,585 tons
4 between Florence and New Orleans	1,617 tons
4 in St. Louis local trade	1,001 tons
7 in local cotton trade	2,016 tons
River "tramps" and unclassified	23,206 tons

It may be noted that in all the years of the development of the Mississippi shipping, there was comparatively little increase in the size of the individual boats. The "Vesuvius," built in 1814, was 480 tons burthen, 160 feet long, 28.6 feet beam, and drew from five to six feet. The biggest boats of later years were but little larger.

THE MISSISSIPPI PILOT

The aristocrat of the Mississippi River steamboat was the pilot. To him all men deferred. So far as the river service furnished a parallel to the autocratic authority of the sea-going captain or master, he was it. All matters pertaining to the navigation of the boat were in his domain, and right zealously he guarded his authority and his dignity. The captain might determine such trivial matters as hiring or discharging men, buying fuel, or contracting for freight; the clerk might lord it over the passengers, and the mate domineer over the black roustabouts; but the pilot moved along in a sort of isolated grandeur, the true monarch of all he surveyed. If, in his judgment the course of wisdom was to tie up to an old sycamore tree on the bank and remain motionless all night, the boat tied up. The grumblings of passengers and the disapproval of the captain availed naught, nor did the captain often venture upon either criticism or suggestion to the lordly pilot, who was prone to resent such invasion of his dignity in ways that made trouble. Indeed, during the flush times on the Mississippi, the pilots were a body of men possessing painfully acquired knowledge and skill, and so organized as to protect all the privileges which their attainments should win for them. The ability to "run" the great river from St. Louis to New Orleans was not lightly won, nor, for that matter, easily retained, for the Mississippi is ever a fickle flood, with changing landmarks and shifting channel. In all the great volume of literature bearing on the story of the river, the difficulties of its conquest are nowhere so truly recounted as in Mark Twain's *Life on the Mississippi*, the humorous quality of which does not obscure, but rather enhances its value as a picturesque and truthful story of the old-time pilot's life. The pilot began his work in boyhood as a "cub" to a licensed pilot. His duties ranged from bringing refreshments up to the pilot-house, to holding the wheel when some straight stretch or clear, deep channel offered his master a chance to leave his post for a few minutes. For strain on the memory, his education is comparable only to the Chinese system of liberal culture, which comprehends learning by rote some tens of thousands of verses from the works of Confucius and other philosophers of the far East. Beginning at New Orleans, he had to commit to memory the name and appearance of every point of land, inlet, river or bayou mouth, "cut-off," light, plantation and hamlet on either bank of the river all the way to St. Louis. Then, he had to learn them all in their opposite order, quite an independent task, as all of us who learned the multiplication table backward in the days of our youth, will readily understand. These landmarks it was needful for him to recognize by day and by night, through fog or driving rain, when the river was swollen by spring floods, or shrunk in summer to a yellow ribbon meandering through a Sahara of sand. He had need to recognize at a glance the ripple on the water that told of a lurking sand-bar and distinguish it from the almost identical ripple that a brisk breeze would raise. Most perplexing of the perils that beset river navigation are the "snags," or sunken logs that often obstruct the channel. Some towering oak or pine, growing in lusty strength for its half-century or more by the brink of the upper reaches of one of the Mississippi system would, in time, be undermined by the flood and fall into the rushing tide. For weeks it would be rolled along the shallows; its leaves and twigs rotting off, its smaller branches breaking short, until at last, hundreds of miles, perhaps, below the scene of its fall, it would lodge fair in the channel. The gnarled and matted mass of boughs would ordinarily cling like an anchor to the sandy bottom, while the buoyant trunk, as though struggling to break away, would strain upward obliquely to within a few inches of the surface of the muddy water, which—too thick to drink and too thin to plough, as the old saying went—gave no hint of this concealed peril; but the boat running fairly upon it, would have her bows stove in and go quickly to the bottom. After the United States took control of the river and began spending its millions annually in improving it for navigation and protecting the surrounding country against its overflows, "snag-boats" were put on the river, equipped with special machinery for dragging these fallen forest giants from the channel, so that of late years accidents from this cause have been rare. But for many years the riverman's chief reliance was that curious instinct or second sight which enabled the trained pilot to pick his way along the most tortuous channel in the densest fog, or to find the landing of some obscure plantation on a night blacker than the blackest of the roustabouts, who moved lively to the incessant cursing of the mate.

The Mississippi River steamboat of the golden age on the river—the type, indeed, which still persists—was a triumph of adaptability to the service for which she was designed. More than this—she was an egregious architectural sham. She was a success in her light draught, six to eight feet, at most, and in her prodigious carrying capacity. It was said of one of these boats, when skilfully loaded by a gang of practical roustabouts, under the direction of an experienced mate, that the freight she carried, if unloaded on the bank, would make a pile bigger than the boat herself. The hull of the vessel was invariably of wood, broad of beam, light of draught, built "to run on a heavy dew," and with only the rudiments of a keel. Some freight was stowed in the hold, but the engines were not placed there, but on the main deck, built almost flush with the water, and extending unbroken from stem to stern. Often the engines were in pairs, so that the great paddle-wheels could be worked independently of each other. The finest and fastest boats were side-wheelers, but a large wheel at the stern, or two stern wheels, side by side, capable of independent action, were common modes of propulsion. The escape-pipes of the engine were carried high aloft, above the topmost of the tiers of decks, and from each one alternately, when the boat was under way, would burst a gush of steam, with a sound like a dull puff, followed by a prolonged sigh, which could be heard far away beyond the dense forests that bordered the river. A row of posts, always in appearance, too slender for the load they bore, supported the saloon deck some fifteen feet above the main deck. When business was good on the river, the space within was packed tight with freight, leaving barely room enough for passenger gangways, and for the men feeding the roaring furnaces with pine slabs. A great steamer coming down to New Orleans from the cotton country about the Red River, loaded to the water's edge with cotton bales, so that, from the shore, she looked herself like a monster cotton bale, surmounted by tiers of snowy cabins and pouring forth steam and smoke from towering pipes, was a sight long to be remembered. It is a sight, too, that is still common on the lower river, where the business of gathering up the planter's crop and getting it to market has not yet passed wholly into the hands of the railroads.

A DECK LOAD OF COTTON

Above the cargo and the roaring furnaces rose the cabins, two or three tiers, one atop the other, the topmost one extending only about one-third of the length of the boat, and called the "Texas." The main saloon extending the whole length of the boat, save for a bit of open deck at bow and stern, was in comparison with the average house of the time, palatial. On either side it was lined by rows of doors, each opening into a two-berthed stateroom. The decoration was usually ivory white, and on the main panel of each door was an oil painting of some romantic landscape. There Chillon brooded over the placid azure of the lake, there storms broke with jagged lightning in the Andes, there buxom girls trod out the purple grapes of some Italian vineyard. The builders of each new steamer strove to eclipse all earlier ones in the brilliancy of these works of art, and discussion of the relative merits of the paintings on the "Natchez" and those on the "Baton Rouge" came to be the chief theme of art criticism along the river. Bright crimson carpet usually covered the floor of the long, tunnel-like cabin. Down the center were ranged the tables, about which, thrice a day, the hungry passengers gathered to be fed, while from the ceiling depended chandeliers, from which hung prismatic pendants, tinkling pleasantly as the boat vibrated with the throb of her engines. At one end of the main saloon was the ladies' cabin, discreetly cut off by crimson curtains; at the other, the bar, which, in a period when copious libations of alcoholic drinks were at least as customary for men as the cigar to-day, was usually a rallying point for the male passengers.

Far up above the yellow river, perched on top of the "Texas," or topmost tier of cabins, was the pilot-house, that honorable eminence of glass and painted wood which it was the ambition of every boy along the river some day to occupy. This was a great square box, walled in mainly with glass. Square across the front of it rose the huge wheel, eight feet in diameter, sometimes half-sunken beneath the floor, so that the pilot, in moments of stress, might not only grip it with his hands, but stand on its spokes, as well. Easy chairs and a long bench made up the furniture of this sacred apartment. In front of it rose the two towering iron chimneys, joined, near the top by an iron grating that usually carried some gaudily colored or gilded device indicative of the line to which the boat belonged. Amidships, and aft of the pilot-house, rose the two escape pipes, from which the hoarse, prolonged s-o-o-ugh of the high pressure exhaust burst at half-minute intervals, carrying to listeners miles away, the news that a boat was coming.

All this edifice above the hull of the boat, was of the flimsiest construction, built of pine scantling, liberally decorated with scroll-saw work, and lavishly covered with paint mixed with linseed oil. Beneath it were two, four, or six roaring furnaces fed with rich pitch-pine, and open on every side to drafts and gusts. From the top of the great chimneys poured volcanic showers of sparks, deluging the inflammable pile with a fiery rain. The marvel is not that every year saw its quotum of steamers burned to the water's edge, but, rather, that the quota were proportionately so small.

At midnight this apparent inflammability was even more striking. Lights shone from the windows of the long row of cabins, and wherever there was a chink, or a bit of glass, or a latticed blind, the radiance streamed forth as though within were a great mass of fire, struggling, in every way, to escape. Below, the boiler deck was dully illumined by smoky lanterns; but when one of the great doors of the roaring furnace was thrown open, that the half-naked black firemen might throw in more pitch-pine slabs, there shone forth such a fiery glare, that the boat and the machinery—working in the open, and plain to view—seemed wrapped in a Vesuvius of flame, and the sturdy stokers and lounging roustabouts looked like the fiends in a fiery inferno. The danger was not merely apparent, but very real. During the early days of steamboating, fires and boiler explosions were of frequent occurrence. A river boat, once ablaze, could never be saved, and the one hope for the passengers was that it might be beached before the flames drove them overboard. The endeavor to do this brought out some examples of magnificent heroism among captains, pilots, and engineers, who, time and again, stood manfully at their posts, though scorched by flames, and cut off from any hope of escape, until the boat's prow was thrust well into the bank, and the passengers were all saved. The pilots, in the presence of such disaster, were in the sorest straits, and were, moreover, the ones of the boat's company upon whom most depended the fate of those on board. Perched at the very top of a large tinder-box, all avenues of escape except a direct plunge overboard were quickly closed to them. If they left the wheel the current would inevitably swing the boat's head downstream, and she would drift, aimlessly, a flaming funeral pyre for all on board. Many a pilot stood, with clenched teeth, and eyes firm set upon the distant shore, while the fire roared below and behind him, and the terrified passengers edged further and further forward as the flames pressed their way toward the bow, until at last came the grinding sound under the hull, and the sudden shock that told of shoal water and safety. Then, those on the lower deck might drop over the side, or swarm along the windward gangplank to safety, but the pilot too often was hemmed in by the flames, and perished with his vessel.

FEEDING THE FURNACE

In the year 1840 alone there were 109 steamboat disasters chronicled, with a loss of fifty-nine vessels and 205 lives. The high-pressure boilers used on the river, cheaply built, and for many years not subjected to any official inspection, contributed more than their share to the list of accidents. Boiler explosions were so common as to be reckoned upon every time a voyage was begun. Passengers were advised to secure staterooms aft when possible, as the forward part of the boat was the more apt to be shattered if the boiler "went up." Every river town had its citizens who had survived an explosion, and the stock form into which to put the humorous quip or story of the time was to have it told by the clerk going up as he met the captain in the air coming down, with the débris of the boat flying all about them. As the river boats improved in character, disasters of this sort became less frequent, and the United States, by establishing a rigid system of boiler inspection, and compelling engineers to undergo a searching examination into their fitness before receiving a license, has done much to guard against them. Yet to-day, we hear all too frequently of river steamers blown to bits, and all on board lost, though it is a form of disaster almost unknown on Eastern waters where crowded steamboats ply the Sound, the Hudson, the Connecticut, and the Potomac, year after year, with never a disaster. The cheaper material of Western boats has something to do with this difference, but a certain happy-go-lucky, devil-may-care spirit, which has characterized the Western riverman since the days of the broadhorns, is chiefly responsible. Most often an explosion is the result of gross carelessness—a sleepy engineer, and a safety-valve "out of kilter," as too many of them often are, have killed their hundreds on the Western rivers. Sometimes, however, the almost criminal rashness, of which captains were guilty, in a mad rush for a little cheap glory, ended in a deafening crash, the annihilation of a good boat, and the death of scores of her people by drowning, or the awful torture of inhaling scalding steam. Rivalry between the different boats was fierce, and now and then at the sight of a competitor making for a landing where freight and passengers awaited the first boat to land her gangplank, the alert captain would not unnaturally take some risks to get there first. Those were the moments that resulted in methods in the engine room picturesquely described as "feeding the fires with fat bacon and resin, and having a nigger sit on the safety valve." To such impromptu races might be charged the most terrifying accidents in the history of the river.

But the great races, extending sometimes for more than a thousand miles up the river, and carefully planned for months in advance, were seldom, if ever, marred by an accident. For then every man on both boats was on the alert, from pilot down to fuel passer. The boat was trimmed by guidance of a spirit level until she rode the water at precisely the draft that assured the best speed. Her hull was scraped and oiled, her machinery overhauled, and her fuel carefully selected. Picked men made up her crew, and all the upper works that could be disposed of were landed before the race, in order to decrease air resistance. It was the current pleasantry to describe the captain as shaving off his whiskers lest they catch the breeze, and parting his hair in the middle, that the boat might be the better trimmed. Few passengers were taken, for they could not be relied upon to "trim ship," but would be sure to crowd to one side or the other at a critical moment. Only through freight was shipped—and little of that—for there would be no stops made from starting-point to goal. Of course, neither boat could carry all the fuel—pine-wood slabs—needed for a long voyage, but by careful prearrangement, great "flats" loaded with wood, awaited them at specified points in midstream. The steamers slowed to half-speed, the flats were made fast alongside by cables, and nimble negroes transferred the wood, while the race went on. At every riverside town the wharves and roofs would be black with people, awaiting the two rivals, whose appearance could be foretold almost as exactly as that of a railway train running on schedule time. The firing of rifles and cannon, the blowing of horns, the waving of flags, greeted the racers from the shores by day, and great bonfires saluted them by night. At some of the larger towns they would touch for a moment to throw off mail, or to let a passenger leap ashore. Then every nerve of captain, pilot, and crew was on edge with the effort to tie up and get away first. Up in the pilot-house the great man of the wheel took shrewd advantage of every eddy and back current; out on the guards the humblest roustabout stood ready for a life-risking leap to get the hawser to the dock at the earliest instant. All the operations of the boat had been reduced to an exact science, so that when the crack packets were pitted against each other in a long race, their maneuvers would be as exactly matched in point of time consumed as those of two yachts sailing for the "America's" cup. Side by side, they would steam for hundreds of miles, jockeying all the way for the most favorable course. It was a fact that often such boats were so evenly matched that victory would hang almost entirely on the skill of the pilot, and where of two pilots on one boat one was markedly inferior, his watch at the wheel could be detected by the way the rival boat forged ahead. During the golden days on the river, there were many of these races, but the most famous of them all was that between the "Robert E. Lee" and the "Natchez," in 1870. These boats, the pride of all who lived along the river at that time, raced from New Orleans to St. Louis. At Natchez, 268 miles, they were six minutes apart; at Cairo, 1024 miles, the "Lee" was three hours and thirty-four minutes ahead. She came in winner by six hours and thirty-six minutes, but the officers of the "Natchez" claimed that this was not a fair test of the relative speed of the boats, as they had been delayed by fog and for repairs to machinery for about seven hours.

Spectacular and picturesque was the riverside life of the great Mississippi towns in the steamboat days. Mark Twain has described the scenes along the levee at New Orleans at "steamboat time" in a bit of word-painting, which brings all the rush and bustle, the confusion, turmoil and din, clearly to the eye:

"It was always the custom for boats to leave New Orleans between four and five o'clock in the afternoon. From three o'clock onward, they would be burning resin and pitch-pine (the sign of preparation) and so one had the spectacle of a rank, some two or three miles long, of tall, ascending columns of coal-black smoke, a colonnade which supported a roof of the same smoke, blending together and spreading abroad over the city. Every outward-bound boat had its flag flying at the jack-staff, and sometimes a duplicate on the verge-staff astern. Two or three miles of mates were commanding and swearing with more than usual emphasis. Countless processions of freight, barrels, and boxes, were spinning athwart the levee, and flying aboard the stage-planks. Belated passengers were dodging and skipping among these frantic things, hoping to reach the forecastle companion-way alive, but having their doubts about it. Women with reticules and bandboxes were trying to keep up with husbands freighted with carpet sacks and crying babies, and making a failure of it by losing their heads in the whirl and roar and general distraction. Drays and baggage-vans were clattering hither and thither in a wild hurry, every now and then getting blocked and jammed together, and then, during ten seconds, one could not see them for the profanity, except vaguely and dimly. Every windlass connected with every forehatch from one end of that long array of steamboats to the other, was keeping up a deafening whiz and whir, lowering freight into the hold, and the half-naked crews of perspiring negroes that worked them were roaring such songs as 'De las' sack! De las' sack!!' inspired to unimaginable exaltation by the chaos of turmoil and racket that was driving everybody else mad. By this time the hurricane and boiler decks of the packets would be packed and black with passengers, the last bells would begin to clang all down the line, and then the pow-wows seemed to double. In a moment or two the final warning came, a simultaneous din of Chinese gongs with the cry, 'All dat aint going, please to get ashore,' and, behold, the pow-wow quadrupled. People came swarming ashore, overturning excited stragglers that were trying to swarm aboard. One moment later, a long array of stage-planks was being hauled in, each with its customary latest passenger clinging to the end of it, with teeth, nails, and everything else, and the customary latest procrastinator making a wild spring ashore over his head.

"Now a number of the boats slide backward into the stream, leaving wide gaps in the serried rank of steamers. Citizens crowd on the decks of boats that were not to go, in order to see the sight. Steamer after steamer straightens herself up, gathers all her strength, and presently comes swinging by, under a tremendous head of steam, with flags flying, smoke rolling, and her entire crew of firemen and deck hands (usually swarthy negroes) massed together on the forecastle, the best voice in the lot towering in their midst (being mounted on the capstan) waving his hat or a flag, all roaring in a mighty chorus, while the parting cannons boom, and the multitudinous spectators swing their hats and huzza. Steamer after steamer pulls into the line, and the stately procession goes winging its flight up the river."

Until 1865 the steamboats controlled the transportation business of all the territory drained by the Mississippi and its tributaries. But two causes for their undoing had already begun to work. The long and fiercely-fought war had put a serious check to the navigation of the rivers. For long months the Mississippi was barricaded by the Confederate works at Island Number 10, at New Madrid and at Vicksburg. Even after Grant and Farragut had burst these shackles navigation was attended with danger from guerrillas on the banks and trade was dead. When peace brought the promise of better things, the railroads were there to take advantage of it. From every side they were pushing their way into New Orleans, building roadways across the "trembling prairies," and crossing the water-logged country about the Rigolets on long trestles. They penetrated the cotton country and the mineral country. They paralleled the Ohio, the Tennessee, and the Cumberland, as well as the Father of Waters, and the steamboat lines began to feel the heavy hand of competition. Captains and clerks found it prudent to abate something of their dignity. Instead of shippers pleading for deck-room on the boats, the boats' agents had to do the pleading. Instead of levees crowded with freight awaiting carriage there were broad, empty spaces by the river's bank, while the railroad freight-houses up town held the bales of cotton, the bundles of staves, the hogsheads of sugar, the shingles and lumber. On long hauls the railroads quickly secured all the North and South business, though indeed, the hauling of freight down the river for shipment to Europe was ended for both railroads and steamboats, so far as the products raised north of the Tennessee line was concerned. For a new water route to the sea had been opened and wondrously developed. The Great Lakes were the shortest waterway to the Atlantic, and New York dug its Erie Canal which afforded an outlet—pinched and straitened, it is true, but still an outlet—for the cargoes of the lake schooners and the early steamers of the unsalted seas. Even the commonwealths forming the north bank of the Ohio River turned their faces away from the stream that had started them on the pathway to wealth and greatness, and dug canals to Lake Erie, that their wheat, corn, and other products might reach tidewater by the shortest route. The great cargoes from Cincinnati, St. Louis, and Louisville, began to be legends of the past, and the larger boats were put on routes in Louisiana, or on the Mississippi, from Natchez south, while others were reduced to mere local voyages, gathering up freight from points tributary to St. Louis. The glory of the river faded fast, and the final stroke was dealt it when some man of inventive mind discovered that a little, puffing tug, costing one-tenth as much as a fine steamboat, could push broad acres of flatboats, loaded with coal, lumber, or cotton, down the tortuous stream, and return alone at one-tenth the expense of a heavy steamer. That was the final stroke to the picturesqueness and the romance of river life. The volume of freight carried still grows apace, but the glory of Mississippi steamboat life is gone forever.

CHAPTER IX

THE NEW ENGLAND FISHERIES—THEIR PART IN EFFECTING THE SETTLEMENT OF AMERICA—THEIR RAPID DEVELOPMENT—WIDE EXTENT OF THE TRADE—EFFORT OF LORD NORTH TO DESTROY IT—THE FISHERMEN IN THE REVOLUTION—EFFORTS TO ENCOURAGE THE INDUSTRY—ITS PART IN POLITICS AND DIPLOMACY—THE FISHING BANKS—TYPES OF BOATS—GROWTH OF THE FISHING COMMUNITIES—FARMERS AND SAILORS BY TURNS—THE EDUCATION OF THE FISHERMEN—METHODS OF TAKING MACKEREL—THE SEINE AND THE TRAWL—SCANT PROFITS OF THE INDUSTRY—PERILS OF THE BANKS—SOME PERSONAL EXPERIENCES—THE FOG AND THE FAST LINERS—THE TRIBUTE OF HUMAN LIFE.

The summer yachtsman whiling away an idle month in cruises up and down that New England coast which, once stern and rock-bound, has come to be the smiling home of midsummer pleasures, encounters at each little port into which he may run, moldering and decrepit wharves, crowned with weatherbeaten and leaky structures, waterside streets lined with shingled fish-houses in an advanced stage of decay, and acres of those low platforms known as flakes, on which at an earlier day the product of the New England fisheries was spread out to dry in the sun, but which now are rapidly disintegrating and mingling again with the soil from which the wood of their structures sprung. Every harbor on the New England coast, from New Bedford around to the Canadian line, bears these dumb memorials to the gradual decadence of what was once our foremost national industry. For the fisheries which once nursed for us a school of the hardiest seamen, which aroused the jealousy of England and France, which built up our seaport towns, and carried our flag to the furthest corners of the globe, and which in the records both of diplomacy and war fill a prominent place have been for the last twenty years appreciably tending to disappear. Many causes are assigned for this. The growing scarcity of certain kinds of fish, the repeal of encouraging legislation, a change in the taste of certain peoples to whom we shipped large quantities of the finny game, the competition of Canadians and Frenchmen, the great development of the salmon fisheries and salmon canning on the Pacific coast, all have contributed to this decay. It is proper, however, to note that the decadence of the fisheries is to some extent more apparent than real. True, there are fewer towns supported by this industry, fewer boats and men engaged in it; but in part this is due to the fact that the steam fishing boat carrying a large fleet of dories accomplishes in one season with fewer hands eight or ten times the work that the old-fashioned pink or schooner did. And, moreover, as the population of the seaport towns has grown, the apparent prominence of the fishing industry has decreased, as that industry has not grown in proportion to the population. Forty years ago Marblehead and Nantucket were simply fishing villages, and nothing else. To-day the remnants of the fishing industry attract but little attention, in the face of the vastly more profitable and important calling of entertaining the summer visitor. New Bedford has become a great factory town, Lynn and Hull are great centers for the shoemaking industries.

When the Pilgrim Fathers first concluded to make their journey to the New England coast and sought of the English king a charter, they were asked by the thrifty James, what profit might arise. "Fishing," was the answer. Whereupon, according to the narrative of Edward Winslow, the king replied, "So, God have my soul; 'tis an honest trade; 'twas the apostles' own calling." The redoubtable Captain John Smith, making his way to the New England coast from Virginia, happened to drop a fishline over what is known now as George's Bank. The miraculous draught of fishes which followed did not awaken in his mind the same pious reflections to which King James gave expression. Rather was he moved to exultation over the profit which he saw there. "Truly," he said, in a letter to his correspondent in London, "It is a pleasant thing to drop a line and pull up threepence, fivepence, and sixpence as fast as one may haul in." The gallant soldier of fortune was evidently quite awake to the possibilities of profit upon which he had stumbled. Yet, probably even he would have been amazed could he have known that within fifty years not all the land in the colony of Massachusetts Bay, nor in the Providence and Rhode Island plantations produced so much of value as the annual crop the fishermen harvested on the shallow banks off Cape Cod.

As early as 1633 fish began to be exported from Boston, and very shortly thereafter the industry had assumed so important a position that the general court adopted laws for its encouragement, exempting vessels, and stock from taxation, and granting to fishermen immunity from military duty. At the close of the seventeenth century, Massachusetts was exporting over $400,000 worth of fish annually. From that time until well into the middle of the last century the fisheries were so thoroughly the leading industry of Massachusetts that the gilded codfish which crowns the dome of the State House at Boston, only fitly typifies by its prominence above the city the part which its natural prototypes played in building up the commonwealth. In the Revolution and the early wars of the United States, the fishermen suffered severely. Crowded together on the banks, they were easy prey for the British cruisers, who, in time of peace or in time of war, treated them about as they chose, impressing such sailors as seemed useful, and seizing such of their cargo as the whim of the captain of the cruiser might suggest. And even before the colonies had attained the status of a nation, the jealousy and hostility of Great Britain bore heavily on the fortunes of the New England fishermen. It was then, as it has been until the present day, the policy of

Great Britain to build up in every possible way its navy, and to encourage by all imaginable devices the development of a large body of able seamen, by whom the naval vessels might be manned. Accordingly parliament undertook to discourage the American fisherman by hostile legislation, so that a body of deep-sea fishermen might be created claiming English ports for their home. At first the effort was made to prohibit the colonies from exporting fish. The great Roman Catholic countries of France, Spain, and Portugal took by far the greater share of the fish sent out, though the poorer qualities were shipped to the West Indies and there exchanged for sugar and molasses. Against this trade Lord North leveled some of his most offensive measures, proposing bills, indeed, so unjust and tyrannical that outcries were raised against them even in the British House of Lords. To cut off intercourse with the foreign peoples who took the fish of the Yankees by hundreds and thousands of quintals, and gave in return rum, molasses, and bills of exchange on England, to destroy the calling in which every little New England seacoast village was interested above all things, Lord North first proposed to prohibit the colonies trading in fish with any country save the "mother" country, and secondly, to refuse to the people of New England the right to fish on the Great Banks of Newfoundland, thus confining them to the off-shore banks, which already began to show signs of being fished out. Even a hostile parliament was shocked by these measures. Every witness who appeared before the House of Commons testified that they would work irreparable injury to New England, would rob six thousand of her able-bodied men of their means of livelihood, and would drive ten thousand more into other vocations. But the power of the ministry forced the bills through, though twenty-one peers joined in a solemn protest. "We dissent," said they, "because the attempt to coerce, by famine, the whole body of the inhabitants of great and populous provinces, is without example in the history of this, or, perhaps, of any civilized nations." This was in 1775, and the revolution in America had already begun. It was the policy of Lord North to force the colonists to stop their opposition to unjust and offensive laws by imposing upon them other laws more unjust and more offensive still—a sort of homeopathic treatment, not infrequently applied by tyrants, but which seldom proves effective. In this case it aligned the New England fishermen to a man with the Revolutionists. A Tory fisherman would have fared as hard as

> "Old Floyd Ireson for his hard heart
>
> Tarr'd and feathered and carried in a cart,
>
> By the woman of Marblehead."

Nor was this any inconsiderable or puny element which Lord North had deliberately forced into revolt. Massachusetts alone had at the outbreak of the Revolution five hundred fishing vessels, and the town of Marblehead one hundred and fifty sea-going fishing schooners. Gloucester had nearly as many, and all along the coast, from Maine to New York, there were thrifty settlers, farmers and fishermen, by turns, as the season served. New England was preeminently a maritime state. Its people had early discovered that a livelihood could more easily be plucked from the green surges of ocean, white-capped as they sometimes were, than wrested from the green and boulder-crowned hills. Upon the fisheries rested practically all the foreign commerce. They were the foundation upon which were built the superstructure of comfort and even luxury, the evidences of which are impressive even in the richer New England of to-day. Therefore, when the British ministry attacked this calling, it roused against the crown not merely the fisherman and the sailor, but the merchants as well—not only the denizens of the stuffy forecastles of pinks and schooners, but the owners of the fair great houses in Boston and New Bedford. Lord North's edicts stopped some thousands of sturdy sailors from catching cod and selling them to foreign peoples. They accordingly became privateers, and preyed upon British commerce until it became easier for a mackerel to slip through the meshes of a seine than for a British ship to make its usual voyages. The edicts touched the commercial Bostonians in their pockets, and stimulated them to give to the Revolution that countenance and support of the "business classes" which revolutionary movements are apt to lack, and lacking which, are apt to fail.

The war, of course, left the fisheries crippled and almost destroyed. It had been a struggle between the greatest naval power of the world, and a loose coalition of independent colonies, without a navy and without a centralized power to build and maintain one. Massachusetts did, indeed, equip an armed ship to protect her fishermen, but partly because the protection was inadequate, and partly as a result of the superior attractions of privateering, the fishing boats were gradually laid up, until scarcely enough remained in commission to supply the demands of the home merchant for fish. Where there had been prosperity and bustle about wharves, and fish-houses, there succeeded idleness and squalor. Shipbuilding was prostrate, commerce was dead. The sailors returned to the farms, shipped on the privateers, or went into Washington's army. But when peace was declared, they flocked to their boats, and began to rebuild their shattered industry. Marblehead, which went into the war with 12,000 tons of shipping, came out with 1500. Her able-bodied male citizens had decreased in numbers from 1200 to 500. Six hundred of her sons, used to hauling the seine and baiting the trawl, were in British prisons. How many from this and other fishing ports were pressed against their will into service on British men-of-war, history has no figures to show; but there were hundreds. Yet, prostrate as the industry was, it quickly revived, and soon again attained those noble proportions that had enabled

Edmund Burke to say of it, in defending the colonies before the House of Commons:

"No ocean but what is vexed with their fisheries; no climate that is not witness of their toils. Neither the perseverance of Holland, nor the activity of France, nor the dextrous and firm sagacity of English enterprise ever carried this perilous mode of hardy enterprise to the extent to which it has been pushed by this recent people—a people who are still, as it were, in the gristle, and not yet hardened into the bone of manhood."

In 1789, immediately upon the formation of the Government under which we now live, the system of giving bounties to the deep-sea fishermen was inaugurated and was continued down to the middle of the last century, when a treaty with England led to its discontinuance. The wisest statesmen and publicists differ sharply concerning the effect of bounties and special governmental favors, like tariffs and rebates, upon the favored industry, and so, as long as the fishing bounty was continued, its needfulness was sharply questioned by one school, while ever since its withdrawal the opposing school has ascribed to that act all the later ills of the industry. Indeed, as this chapter is being written, a subsidy measure before Congress for the encouragement of American shipping, contains a proviso for a direct payment from the national treasury to fishing vessels, proportioned to their size and the numbers of their crews. It is not my purpose to discuss the merits, either of the measure now pending, or of the many which have, from time to time, encouraged or depressed our fishermen. It would be hard, however, for any one to read the history of the fisheries without being impressed by the fact that the hardy and gallant men who have risked their lives in this most arduous of pursuits, have suffered from too much government, often being sorely injured by a measure intended solely for their good, as in the case of the Treaty of 1818. That instrument was negotiated for the purpose of maintaining the rights of American fishermen on the banks off Newfoundland, Labrador, and Nova Scotia. The American commissioners failed to insist upon the right of the fishermen to land for bait, and this omission, together with an ambiguity in defining the "three-mile limit," enabled the British government to harass, harry, and even confiscate American fishermen for years. American fleets were sent into the disputed waters, and two nations were brought to the point of war over the question which should control the taking of fish in waters that belonged to neither, and that held more than enough for all peoples. To settle the dispute the United States finally entered into another treaty which secured the fishermen the rights ignored in the treaty of 1818, but threw American markets open to Canadian fishermen. This the men of Gloucester and Marblehead, nurtured in the school of protection, declared made their last state worse than the first. So the tinkering of statutes and treaties went on, even to the present day, the fisheries languishing meanwhile, not in our country alone, but in all engaged in the effort to get special privileges on the fishing grounds. Whenever man tries thus to monopolize, by sharp practise or exclusive laws, the bounty which God has provided in abundance for all, the end is confusion, distress, disaster, and too often war.

But the story of what the politicians, and those postgraduates of politics, the statesmen, have done for and against the fishermen of New England, is not that which I have to tell. Rather, it is my purpose to tell something of the lives of the fishermen, the style of their vessels, the portions of the rolling Atlantic which they visit in search of their prey, their dire perils, their rough pleasures, and their puny profits. First, then, as to their prey, and its haunts.

The New England fishermen, in the main, seek three sorts of fish—the mackerel, the cod, and the halibut. These they find on the shallow banks which border the coast from the southern end of Delaware to the very entrance of Baffin's Bay. The mackerel is a summer fish, coming and going with the regularity of the equinoxes themselves. Early in March, they appear off the coast, and all summer work their way northward, until, in early November, they disappear off the coast of Labrador, as suddenly as though some titanic seine had swept the ocean clear of them. What becomes of the mackerel in winter, neither the inquisitive fisherman nor the investigating scientist has ever been able to determine. They do not, like migratory birds, reappear in more temperate southern climes, but vanish utterly from sight. Eight months, therefore, is the term of the mackerel fishing, and the men engaged in it escape the bitterest rigors of the winter fisheries on the Newfoundland Banks, where the cod is taken from January to January. Yet it has dangers of its own—dangers of a sort that, to the sailor, are more menacing than the icebergs or even the swift-rushing ocean liners of the Great Banks. For mackerel fishing is pursued close in shore, in shallow water, where the sand lies a scant two fathoms below the surface, and a north-east wind will, in a few minutes, raise, a roaring sea that will pound the stoutest vessel to bits against the bottom. With plenty of sea-room, and water enough under the keel, the sailor cares little for wind or waves; but in the shallows, with the beach only a few miles to the leeward, and the breakers showing white through the darkness, like the fangs of a beast of prey, the captain of a fishing schooner on George's banks has need of every resource of the sailor, if he is to beat his way off, and not feed the fishes that he came to take. Nowhere is the barometer watched more carefully than on the boats cruising about on George's. When its warning column falls, the whole fleet makes for the open sea, however good the fishing may be. But, with all possible caution, the losses are so many that George's, early in its history, came to have the ghoulish nickname of "Dead Men's Bank."

North of George's Bank—which lies directly east of Cape Cod—are found, in order, Brown's Bank, La Have, Western Bank—in the center of which lies Sable Island, famed as an ocean graveyard, whose shifting sands are as thickly strewn with the bleaching ribs of stout ships as an old green churchyard is set with mossy marbles—St. Peter's Bank, and the Grand Bank of Newfoundland. All of these lie further out to sea than George's, and are tenanted only by cod and halibut, though in the waters near the shore the fishermen pursue the mackerel, the herring—which, in cottonseed oil masquerades as American sardines—and the menhaden, used chiefly for fertilizer. The boats used in the fisheries are virtually of the same model, whatever the fish they may seek—except in the case of the menhaden fishery, which more and more is being prosecuted in slow-going steamers, with machines for hauling seines, and trawl nets. But the typical fishing boat engaged in the food fisheries is a trim, swift schooner, built almost on the lines of a yacht, and modeled after a type designed by Edward Burgess, one of New England's most famous yacht designers. Seaworthy and speedy both are these fishing boats of to-day, fit almost to sail for the "America's" cup, modeled, as they are, from a craft built by the designer of a successful cup defender. That the fishermen ply their calling in vessels so perfectly fitted to their needs is due to a notable exhibition of common sense and enterprise on the part of the United States Fish Commission. Some years ago almost anything that would float was thought good enough for the bank fishermen. In the earliest days of the industry, small sloops were used. These gave way to the "Chebacco boat," a boat taking its name from the town of Chebacco, Massachusetts, where its rig was first tested. This was a fifteen to twenty ton boat almost as sharp at the stern as in the bow, carrying two masts, both cat-rigged. A perfect marvel of crankiness a boat so rigged would seem; but the New England seamen became so expert in handling them that they took them to all of the fishing banks, and even made cruises to the West Indies with cargoes of fish, bringing back molasses and rum. A development of the Chebacco boat was the pink, differing only in its rig, which was of the schooner model. But in time the regular schooner crowded out all other types of fishing vessels. In 1882, the members of the Fish Commission, studying the frightful record of wrecks and drownings among the Gloucester and Marblehead fishermen, reached the conclusion that an improved model fishing boat might be the means of saving scores of lives. The old model was seen to be too heavily rigged, with too square a counter, and insufficient draught. Accordingly, a model boat, the "Grampus," was designed, the style of which has been pretty generally followed in the fishing fleet.

ON THE BANKS.

Such a typical craft is a schooner of about eighty tons, clean-cut about the bows, and with a long overhang at the stern that would give her a rakish, yacht-like air, except for the evidences of her trade, with which her deck is piled. Her hull is of the cutter model, sharp and deep, affording ample storage room. She has a cabin aft, and a roomy forecastle, though such are the democratic conditions of the fishing trade that part of the crew bunks aft with the skipper. The galley, a little box of a place, is directly abaft the foremast, and back of it to the cabin, are the fishbins for storing fish, after they are cleaned and salted or iced. Nowadays, when the great cities, within a few hours' sail of the banks, offer a quick market for fresh fish, many of the fishing boats bring in their catch alive—a deep well, always filled with sea-water, taking the place of the fishbins. The deck, forward of the trunk cabin, is flush, and provided with "knockdown" partitions, so that hundreds of flapping fish may be confined to any desired portion. Amidships of the bankers rises a pile of five or six dories, the presence of which tells the story of the schooner's purpose, for fishing on the Grand Banks for cod is mainly done with trawls which must be tended from dories—a method which has resulted in countless cruel tragedies.

The lives of the men who go down to the sea in ships are always full of romance, the literary value of which has been fully exploited by such writers of sea stories as Cooper and Clark Russell. But the romance of the typical sailor's life is that which grows out of a ceaseless struggle with the winds and waves, out of world-wide wanderings, and encounters with savages and pirates. It is the romance which makes up melodrama, rather than that of the normal life. The early New England fishermen, however, were something more than vagrants on the surface of the seas. In their lives were often combined the peaceful vocations of the farmer or woodsman, with the adventurous calling of the sailor. For months out of the year, the Maine fisherman would be working in the forests, felling great trees, guiding the tugging ox-teams to the frozen rivers, which with spring would float the timber down to tidewater. When winter's grip was loosened, he, like the sturdy logs his axe had shaped, would find his way to where the air was full of salt, and the owners of pinks and schooners were painting their craft, running over the rigging, and bargaining with the outfitters for stores for the spring cruise. From Massachusetts and Rhode Island farms men would flock to the little ports, leaving behind the wife and younger boys to take care of the homestead, until the husband and father returned from the banks in the fall, with his summer's earnings. His luck at fishing, her luck with corn and calves and pigs, determined the scale of the winter's living. Some of the fishermen were not only farmers, as well, but ship-builders and ship-owners, too. If the farm happened to front on some little cove, the frame of a schooner would be set up there on the beach, and all winter long the fisherman-farmer-builder would work away with adze and saw and hammer, putting together the stout hull that would defend him in time against the shock of the north-east sea. His own forest land supplied the oak trees, keelson, ribs, and stem. The neighboring sawmill shaped his planks. One lucky cruise as a hand on a fishing boat owned by a friend would earn him enough to pay for the paint and cordage. With Yankee ingenuity he shaped the iron work at his own forge—evading in its time the stupid British law that forbade the colonists to make nails or bolts. Two winters' labor would often give the thrifty builder a staunch boat of his own, to be christened the "Polly Ann," or the "Mary Jane"—more loyal to family ties than to poetic euphony were the Yankee fishermen—with which he would drive into the teeth of the north-east gale, breaking through the waves as calmly as in early spring at home he forced his plough through the stubble.

There was, too, in those early days of the fisheries, a certain patriarchal relation maintained between owner and crew that finds no parallel in modern times. The first step upward of the fisherman was to the quarter-deck. As captain, he had a larger responsibility, and received a somewhat larger share of the catch, than any of his crew. Then, if thrifty, or if possessed of a shipyard at home, such as I have described, he soon became an owner. In time, perhaps, he would add one or two schooners to his fleet, and then stay ashore as owner and outfitter, sending out his boats on shares. Fishermen who had attained to this dignity, built those fine, old, great houses, which we see on the water-front in some parts of New England—square, simple, shingled to the ground, a deck perched on the ridge-pole of the hipped roof, the frame built of oak shaped like a ship's timbers, with axe and adze. The lawns before the houses sloped down to the water where, in the days of the old prosperity, the owner's schooner might be seen, resting lightly at anchor, or tied up to one of the long, frail wharves, discharging cargo—wharves black and rotting now, and long unused to the sailor's cheery cry. There, too, would be the flakes for drying fish, the houses on the wharves for storing supplies, and the packed product, and the little store in which the outfitter kept the simple stock of necessaries from which all who shipped on his fleet were welcome to draw for themselves and their families, until their "ship came in." To such a fishing port would flock the men from farm and forest, as the season for mackerel drew nigh. The first order at the store would include a pair of buck (red leather) or rubber boots, ten or fifteen pounds of tobacco, clay pipe, sou'-westers, a jack-knife, and oil-clothes. If the sailor was single, the account would stop there, until his schooner came back to port. If he had a family, a long list of groceries, pork and beans, molasses, coffee, flour, and coarse cloth, would be bought on credit, for the folks at home. It came about naturally that these folks preferred to be near the store at which the family had credit, and so the sailors would, in time, buy little plots of land in the neighborhood, and build thereon their snug shingled cottages. So sprung up the fishing villages of New England.

The boys who grew up in these villages were able to swim as soon as they could walk; rowed and sailed boats before they could guide a plow; could give the location of every bank, the sort of fish that frequented it, and the season for taking them. They could name every rope and clew, every brace and stay on a pink or Chebacco boat before they reached words of two syllables in Webster's blue-backed spelling-book; the mysteries of trawls and handlines, of baits and hooks were unraveled to them while still in the nursery, and the songs that lulled them to sleep were often doleful ditties of castaways on George's Bank. Often they were shipped as early as their tenth year, going as a rule in schooners owned or commanded by relatives. It was no easy life that the youngster entered upon when first he attained the dignity of being a "cut-tail," but such as it was, it was the life he had looked forward to ever since he was old enough to consider the future. He lived in a little forecastle, heated by a stuffy stove, which it was his business to keep supplied with fuel. The bunks on either side held rough men, not over nice of language or of act, smoking and playing cards through most of their hours of leisure. From time immemorial it has been a maxim of the forecastle that the way to educate a boy is to "harden" him, and the hardening process has usually taken the form of persistent brutality of usage—the rope's end, the heavy hand, the hard-flung boot followed swift upon transgression of the laws or customs of ship or forecastle. The "cut-tail" was everybody's drudge, yet gloried in it, and a boy of Gloucester or Marblehead, who had lived his twelve years without at least one voyage to his credit, was in as sorry a state among his fellow urchins as a "Little Lord Fauntleroy" would be in the company of Tom Sawyer and Huckleberry Finn.

The intimacies of the village streets were continued on the ocean. Fish supplanted marbles as objects of prime importance in the urchin's mind. The smallest fishing village would have two or three boats out on the banks, and the larger town several hundred. Between the crews of these vessels existed always the keenest rivalry, which had abundant opportunity for its exhibition, since the conditions of the fishery were such that the schooners cruised for weeks, perhaps, in fleets of several hundred. Every maneuver was made under the eyes of the whole fleet, and each captain and sailor felt that among the critics were probably some of his near neighbors at home. Charles Nordhoff, who followed a youth spent at sea with a long life of honorable and brilliant activity in journalism, describes the watchfulness of the fleet as he had often seen it:

"The fleet is the aggregate of all the vessels engaged in the mackerel fishery. Experience has taught fishermen that the surest way to find mackerel is to cruise in one vast body, whose line of search will then extend over an area of many miles. When, as sometimes happens, a single vessel falls in with a large 'school,' the catch is, of course, much greater. But vessels cruising separately or in small squads are much less likely to fall in with fish than is the large fleet. 'The fleet' is therefore the aim of every mackerel fisherman. The best vessels generally maintain a position to the windward. Mackerel mostly work to windward slowly, and those vessels furthest to windward in the fleet are therefore most likely to fall in with fish first, while from their position they can quickly run down should mackerel be raised to leeward.

"Thus, in a collection of from six hundred to a thousand vessels, cruising in one vast body, and spreading over many miles of water, is kept up a constant, though silent and imperceptible communication, by means of incessant watching with good spy-glasses. This is so thorough that a vessel at one end of the fleet cannot have mackerel 'alongside,' technically speaking, five minutes, before every vessel in a circle, the diameter of which may be ten miles, will be aware of the fact, and every man of the ten thousand composing their crews will be engaged in spreading to the wind every available stitch of canvas to force each little bark as quickly as possible into close proximity to the coveted prize."

To come upon the mackerel fleet suddenly, perhaps with the lifting of the fog's gray curtain, or just as the faint dawn above the tossing horizon line to the east began to drive away the dark, was a sight to stir the blood of a lad born to the sea. Sometimes nearly a thousand vessels would be huddled together in a space hardly more than a mile square. At night, their red and green lights would swing rhythmically up and down as the little craft were tossed by the long rollers of old Atlantic, in whose black bosom the gay colors were reflected in subdued hues. From this floating city, with a population of perhaps ten thousand souls, no sound arises except the occasional roar of a breaking swell, the creaking of cordage, and the "chug-chug" of the vessel's bows as they drop into the trough of the sea. All sails are furled, the bare poles showing black against the starlit sky, and, with one man on watch on the deck, each drifts idly before the breeze. Below, in stuffy cabins and fetid forecastles, the men are sleeping the deep and dreamless sleep that hard work in the open air brings as one of its rewards. All is as quiet as though a mystic spell were laid on all the fleet. But when the sky to the eastward begins to turn gray, signs of life reappear. Here and there in the fleet a sail will be seen climbing jerkily to the masthead, and hoarse voices sound across the waters. It is only a minute or two after the first evidence of activity before the whole fleet is tensely active. Blocks and cordage are creaking, captains and mates shouting. Where there was a forest of bare poles are soon hundreds of jibs and mainsails, rosy in the first rays of the rising sun. The schooners that have been drifting idly, are, as by magic, under weigh, cutting across each other's bows, slipping out of menacing entanglements, avoiding collisions by a series of nautical miracles. From a thousand galleys rise a thousand slender wreaths of smoke, and the odors of coffee and of the bean dear to New England fishermen, mingle with the saline zephyrs of the sea. The fleet is awake.

They who have sailed with the fleet say that one of the marvels of the fisherman's mind is the unerring skill with which he will identify vessels in the distant fleet, To the landsman all are alike—a group of somewhat dingy schooners, not over trig, and apt to be in need of paint. But the trained fisherman, pursing his eyes against the sun's glitter on the waves, points them out one by one, with names, port-of-hail, name of captain, and bits of gossip about the craft. As the mountaineer identifies the most distant peak, or the plainsman picks his way by the trail indistinguishable to the untrained eye, so the fisherman, raised from boyhood among the vessels that make up the fleet, finds in each characteristics so striking, so individual, as to identify the vessel displaying them as far as a keen eye can reach.

"THE BOYS MARKED THEIR FISH BY CUTTING OFF THEIR TAILS"

The fishing schooners, like the whalers, were managed upon principles of profit-sharing. The methods of dividing the proceeds of the catch differed, but in no sense did the wage system exist, except for one man on board— the cook, who was paid from $40 to $60 a month, besides being allowed to fish in return for caring for the vessel when all the men were out in dories. Sometimes the gross catch of the boat was divided into two parts, the owners who outfitted the boat, supplying all provisions, equipment, and salt, taking one part, the other being divided among the fishermen in proportion to the catch of each. Every fish caught was carefully tallied, the customary method being to cut the tongues, which at the lose of the day's work were counted by the captain, and each man's catch credited. The boys, of whom each schooner carried one or two, marked their fish by cutting off the tails, wherefore these hardy urchins, who generally took the sea at the age of ten, were called "cut-tails." The captain, for his more responsible part in the management of the boat, was not always expected to keep tally of his fish, but was allowed an average catch, plus from three to five per cent. of the gross value of the cargo. Not infrequently the captain was owner of the boat, and his crew, thrifty neighbors of his, owning their own houses by the waterside, and able to outfit the craft and provide for the sustenance of their wives and children at home without calling upon the capitalist for aid. In such a case, the whole value of the catch was divided among the men who made it. At best, these shares were not of a sort to open the doors of a financial paradise to the men. The fisheries have always afforded impressive illustrations of the iron rule of the business world that the more arduous and more dangerous an occupation is, the less it pays. It was for the merest pittance that the fishermen risked their lives, and those who had families at home drawing their weekly provender from the outfitter were lucky if, at the end of the cruise they found themselves with the bill at the store paid, and a few dollars over for necessaries during the winter. In 1799, when the spokesmen of the fishery interests appeared before Congress to plead for aid, they brought papers from the town of Marblehead showing that the average earnings of the fishing vessels hailing from that port were, in 1787, $483; in 1788, $456; and in 1789, $273. The expenses of each vessel averaged $275. In the best of the three years, then, there was a scant $200 to be divided among the captain, the crew, and the owner. This was, of course, one of the leanest of the lean years that the fishermen encountered; but with all the encouragement in the way of bounties and protected markets that Congress could give them, they never were able to earn in a life, as much as a successful promoter of trusts nowadays will make in half an hour. The census figures of 1890—the latest complete figures on occupations and earnings—give the total value of American fisheries as $44,500,000; the number of men employed in them, 132,000, and the average earnings $337 a man. The New England fisheries alone were then valued at $14,270,000. In the gross total of the value of American fisheries are included many methods foreign to the subject of this book, as for example, the system of fishing from shore with pound nets, the salmon fisheries of the Columbia River, and the fisheries of the Great Lakes.

Mackerel are taken both with the hook and in nets—taken in such prodigious numbers that the dories which go out to draw the seine are loaded until their gunwales are almost flush with the sea, and each haul seems indeed a miraculous draught of fishes. It is the safest and pleasantest form of fishing known to the New Englander, for its season is in summer only; the most frequented banks are out of the foggy latitude, and the habit of the fish of going about in monster schools keeps the fishing fleet together, conducing thus to safety and sociability both. In one respect, too, it is the most picturesque form of fishing. The mackerel is not unlike his enemy, man, in his curiosity concerning the significance of a bright light in the dark. Shrewd shopkeepers, who are after gudgeons of the human sort, have worked on this failing of the human family so that by night some of our city streets blaze with every variety of electric fire. The mackerel fisherman gets after his prey in much the same fashion. When at night the lookout catches sight of the phosphorescent gleams in the water that tells of the restless activity beneath of a great school of fish the schooner is headed straightway for the spot. Perhaps forty or fifty other schooners will be turning their prows the same way, their red and green lights glimmering through the black night on either side, the white waves under the bows showing faintly, and the creaking of the cordage sounding over the waters. It is a race for first chance at the school, and a race conducted with all the dash and desperation of a steeple-chase. The skipper of each craft is at his own helm, roaring out orders, and eagerly watchful of the lights of his encroaching neighbors. With the schooner heeled over to leeward, and rushing along through the blackness, the boats are launched, and the men tumble over the side into them, until perhaps the cook, the boy, and the skipper are alone on deck. One big boat, propelled by ten stout oarsmen, carries the seine, and with one dory is towed astern the schooner until the school is overhauled, then casts off and leaps through the water under the vigorous tugs of its oarsmen. In the stern a man stands throwing over the seine by armsful. It is the plan of campaign for the long boat and the dory, each carrying one end of the net, to make a circuit of the school, and envelope as much of it as possible in the folds of the seine. Perhaps at one time boats from twenty or thirty schooners will be undertaking the same task, their torches blazing, their helmsmen shouting, the oars tossing phosphorescent spray into the air. In and out among the boats the schooners pick their way—a delicate task, for each skipper wishes to keep as near as possible to his men, yet must run over neither boats or nets belonging to his rival. Wonderfully expert helmsmen they become after years of this sort of work—more trying to the nerves and exacting quite as much skill as the "jockeying" for place at the start of an international yacht race.

When the slow task of drawing together the ends of the seine until the fish are fairly enclosed in a sort of marine canal, a signal brings the schooner down to the side of the boats. The mackerel are fairly trapped, but the glare of the torches blinds them to their situation, and they would scarcely escape if they could. One side of the net is taken up on the schooner's deck, and there clamped firmly, the fish thus lying in the bunt, or pocket between the schooners, and the two boats which lie off eight or ten feet, rising and falling with the sea. There, huddled together in the shallow water, growing ever shallower as the net is raised, the shining fish, hundreds and thousands of them, bushels, barrels, hogsheads of them, flash and flap, as the men prepare to swing them aboard in the dip net. This great pocket of cord, fit to hold perhaps a bushel or more, is swung from the boom above, and lowered into the midst of the catch. Two men in the boat seize its iron rim, and with a twist and shove scoop it full of mackerel. "Yo-heave-oh" sing out the men at the halliards, and the net rises into the air, and swings over the deck of the schooner. Two men perched on the rail seize the collar and, turning it inside out, drop the whole finny load upon the deck. "Fine, fat, fi-i-ish!" cry out the crew in unison, and the net dips back again into the corral for another load. So, by the light of smoky torches, fastened to the rigging, the work goes on, the men singing and shouting, the tackle creaking, the waves splashing, the wind singing in the shrouds, the boat's bow bumping dully on the waves as she falls. To all these sounds of the sea comes soon to be added one that is peculiar to the banks, a sound rising from the deck of the vessel, a multitude of little taps, rhythmical, muffled, soft as though a corps of clog-dancers were dancing a lively jig in rubber-soled shoes. It is the dance of death of the hapless mackerel. All about the deck they flap and beat their little lives away. Scales fly in every direction, and the rigging, almost to the masthead, is plastered with them.

When the deck is nearly full—and sometimes a single haul of the seine will more than fill it twice—the labor of dipping is interrupted and all hands turn to with a will to dress and pack the fish. Not pretty work, this, and as little pleasing to perform. Barrels, boards, and sharp knives are in requisition. Torches are set up about the deck. The men divide up into gangs of four each and group themselves about the "keelers," or square, shallow boxes into which the fish to be dressed are bailed from the deck. Two men in each gang are "splitters"; two "gibbers." The first, with a dextrous slash of a sharp knife splits the fish down the back, and throws it to the "gibber," who, with a twist of his thumb—armed with a mitt—extracts the entrails and throws the fish into a barrel of brine. By long practise the men become exceedingly expert in the work, and rivalry among the gangs keeps the pace of all up to the highest possible point. All through the night they work until the deck is cleaned of fish, and slimy with blood and scales. The men, themselves, are ghastly, besmeared as they are from top to toe with the gore of the mackerel. From time to time, full barrels are rolled away, and lowered into the hold, and fresh fish raised from the slowly emptying seine alongside. Until the last fish has been sliced, cleaned, plunged into brine, and packed away there can be little respite from the muscle grinding work. From time to time, the pail of tepid water is passed about; once at least during the night, the cook goes from gang to gang with steaming coffee, and now and then some man whose wrist is wearied beyond endurance, knocks off, and with contortions of pain, rubs his arm from wrist to elbow. But save for these momentary interruptions, there is little break in the work. Meanwhile the boat is plunging along through the water, the helm lashed or in beckets, and the skipper hard at work with a knife or gibbing mitt. A score of other boats in a radius of half a mile or so, will be in like case, so there is always danger of collision. Many narrow escapes and not a few accidents have resulted from the practice of cleaning up while under sail.

FISHING FROM THE RAIL

The mackerel, however, is not caught solely in nets, but readily takes that oldest of man's predatory instruments, the hook. To attract them to the side of the vessel, a mixture of clams and little fish called "porgies," ground together in a mill, is thrown into the sea, which, sinking to the depths at which the fish commonly lie, attract them to the surface and among the enticing hooks. Every fisherman handles two lines, and when the fishing is good he is kept busy hauling in and striking off the fish until his arms ache, and the tough skin on his hands is nearly chafed through. Sometimes the hooks are baited with bits of clam or porgy, though usually the mackerel, when biting at all, will snap with avidity at a naked hook, if tinned so as to shine in the water. Mr. Nordhoff, whose reminiscences of life on a fishing boat I have already quoted, describes this method of fishing and its results graphically:

"At midnight, when I am called up out of my warm bed to stand an hour's watch, I find the vessel pitching uneasily, and hear the breeze blowing fretfully through the naked rigging. Going on deck, I perceive that both wind and sea have 'got up' since we retired to rest. The sky looks lowering, and the clouds are evidently surcharged with rain. In fine the weather, as my predecessor on watch informs me, bears every sign of an excellent fishday on the morrow. I accordingly grind some bait, sharpen up my hooks once more, see my lines clear, and my heaviest jigs (the technical term for hooks with pewter on them) on the rail ready for use, and at one o'clock return to my comfortable bunk. I am soon again asleep, and dreaming of hearing fire-bells ringing, and seeing men rush to the fire, and just as I see 'the machine' round the corner of the street, am startled out of my propriety, my dream, sleep, and all by the loud cry of 'Fish!'

"I start up desperately in my narrow bunk, bringing my cranium in violent contact with a beam overhead, which has the effect of knocking me flat down in my berth again. After recovering as much consciousness as is necessary to appreciate my position, I roll out of bed, jerk savagely at my boots, and snatching up my cap and pea-jacket, make a rush *at* the companion-way, *up* which I manage to fall in my haste, and then spring into the hold for a strike-barrel.

"And now the mainsail is up, the jib down, and the captain is throwing bait. It is not yet quite light, but we hear other mainsails going up all round us. A cool drizzle makes the morning unmistakably uncomfortable, and we stand around half asleep, with our sore hands in our pockets, wishing we were at home. The skipper, however, is holding his lines over the rail with an air which clearly intimates that the slightest kind of a nibble will be quite sufficient this morning to seal the doom of a mackerel.

"'There, by Jove!' the captain hauls back—'there, I told you so! Skipper's got him—no—aha, captain, you haul back too savagely!'

"With the first movement of the captain's arm, indicating the presence of fish, everybody rushes madly to the rail. Jigs are heard on all sides plashing into the water, and eager hands and arms are stretched at their full length over the side, feeling anxiously for a nibble.

"'Sh—hish—there's something just passed my fly—I felt him,' says an old man standing alongside of me.

"'Yes, and I've got him,' triumphantly shouts out the next man on the other side of him, hauling in as he speaks, a fine mackerel, and striking him off into his barrel in the most approved style.

"Z-z-zip goes my line through and deep into my poor fingers, as a huge mackerel rushes savagely away with what he finds not so great a prize as he thought it was. I get confoundedly flurried, miss stroke half a dozen times in hauling in as many fathoms of line, and at length succeed in landing my first fish safely in my barrel, where he flounders away 'most melodiously,' as my neighbor says.

"And now it is fairly daylight, and the rain, which has been threatening all night, begins to pour down in right earnest. As the heavy drops patter on the sea the fish begin to bite fast and furiously.

"'Shorten up,' says the skipper, and we shorten in our lines to about eight feet from the rail to the hooks, when we can jerk them in just as fast as we can move our hands and arms. 'Keep your lines clear,' is now the word, as the doomed fish slip faster and faster into the barrels standing to receive them. Here is one greedy fellow already casting furtive glances behind him, and calculating in his mind how many fish he will have to lose in the operation of getting his second strike-barrel.

"Now you hear no sound except the steady flip of fish into the barrels. Every face wears an expression of anxious determination; everybody moves as though by springs; every heart beats loud with excitement, and every hand hauls in fish and throws out hooks with a methodical precision, a kind of slow haste, which unites the greatest speed with the utmost security against fouling lines.

"And now the rain increases. We hear jibs rattling down; and glancing up hastily, I am surprised to find our vessel surrounded on all sides by the fleet, which has already become aware that we have got fish alongside. Meantime the wind rises, and the sea struggles against the rain, which is endeavoring with its steady patter to subdue the turmoil of old ocean. We are already on our third barrel each, and still the fish come in as fast as ever, and the business (sport it has ceased to be some time since), continues with vigor undiminished. Thick beads of perspiration chase each other down our faces. Jackets, caps, and even over-shirts, are thrown off, to give more freedom to the limbs that are worked to their utmost.

"'Hillo! Where are the fish?' All gone. Every line is felt eagerly for a bite, but not the faintest nibble is perceptible. The mackerel, which but a moment ago were fairly rushing on board, have in that moment disappeared so completely that not a sign of one is left. The vessel next under our lee holds them a little longer than we, but they finally also disappear from her side. And so on all around us.

"And now we have time to look about us—to compare notes on each other's successes—to straighten our backbones, nearly broken and aching horribly with the constant reaching over; to examine our fingers, cut to pieces and grown sensationless with the perpetual dragging of small lines across them—to—'There, the skipper's got a bite! Here they are again, boys, and big fellows, too!' Everybody rushes once more to the rail, and business commences again, but not at so fast a rate as before. By-and-by there is another cessation, and we hoist our jib and run off a little way, into a new berth.

"While running across, I take the first good look at the state of affairs in general. We lie, as before said, nearly in the center of the whole fleet, which from originally covering an area of perhaps fifteen miles each way, has 'knotted up' into a little space, not above two miles square. In many places, although the sea is tolerably rough, the vessels lie so closely together that one could almost jump from one to the other. The greatest skill and care are necessary on such occasions to keep them apart, and prevent the inevitable consequences of a collision, a general smash-up of masts, booms, bulwarks, etc. Yet a great fish-day like this rarely passes off without some vessel sustaining serious damage. We thread our way among the vessels with as much care and as daintily as a man would walk over ground covered with eggs; and finally get into a berth under the lee of a vessel which seems to hold the fish pretty well. Here we fish away by spells, for they have become 'spirty,' that is, they are capricious, and appear and disappear suddenly."

TRAWLING FROM A DORY

Three causes make the occupation of those fishermen who go for cod and halibut to the Newfoundland Banks extra hazardous—the almost continual fog, the swift steel Atlantic liners always plowing their way at high speed across the fishing grounds, heedless of fog or darkness, and the custom of fishing with trawls which must be tended from dories. The trawl, which is really only an extension of hand-lines, is a French device adopted by American fishermen early in the last century. One long hand-line, supported by floats, is set at some distance from the schooner. From it depend a number of short lines with baited hooks, set at brief intervals. The fisherman, in his dory, goes from one to the other of these lines pulling them in, throwing the fish in the bottom of the boat and rebaiting his hooks. When his dory is full he returns with his load to the schooner—if he can find her.

That is the peril ever present to the minds of the men in the dory—the danger of losing the schooner. On the Banks the sea is always running moderately high, and the great surges, even on the clearest days, will often shut out the dories from the vision of the lookout. The winds and the currents tend to sweep the little fishing-boats away, and though a schooner with five or six dories out hovers about them like a hen guarding her chickens, sailing a triangular beat planned to include all the smaller boats, yet it too often happens that night falls with one boat missing. Then on the schooner all is watchfulness. Cruising slowly about, burning flares and blowing the hoarse fog-horn, those on board search for the missing ones until day dawns or the lost are found. Sometimes day comes in a fog, a dense, dripping, gray curtain, more impenetrable than the blackest night, for through it no flare will shine, and even the sound of the braying horn or tolling bell is so curiously distorted, that it is difficult to tell from what quarter it comes. No one who has not seen a fog on the Banks can quite imagine its dense opaqueness. When it settles down on a large fleet of fishermen, with hundreds of dories out, the peril and perplexity of the skippers are extreme. In one instant after the dull gray curtain falls over the ocean, each vessel is apparently as isolated as though alone on the Banks. A dory forty feet away is invisible. The great fleet of busy schooners, tacking back and forth, watching their boats, is suddenly, obliterated. Hoarse cries, the tooting of horns and the clanging of bells, sound through the misty air, and now and then a ghostly schooner glides by, perhaps scraping the very gunwale and carrying away bits of rail and rigging to the accompaniment of New England profanity. This is the dangerous moment for every one on the Banks, for right through the center of the fishing ground lies the pathway of the great steel ocean steamships plying between England and the United States. Colossal engines force these great masses of steel through sea and fog. Each captain is eager to break a record; each one knows that a reputation for fast trips will make his ship popular and increase his usefulness to the company. In theory he is supposed to slow down in crossing the Banks; in fact his great 12,000-ton ship rushes through at eighteen miles an hour. If she hits a dory and sends two men to their long rest, no one aboard the ocean leviathan will ever know it. If she strikes a schooner and shears through her like a knife through cheese, there will be a slight vibration of the steel fabric, but not enough to alarm the passengers; the lookout will have caught a hasty glimpse of a ghostly craft, and heard plaintive cries for help, then the fog shuts down on all, like the curtain on the last act of a tragedy. Even if the great steamship were stopped at once, her momentum would carry her a mile beyond the spot before a boat could be lowered, and then it would be almost impossible to find the floating wreckage in the fog. So, usually, the steamships press on with unchecked speed, their officers perhaps breathing a sigh of pity for the victims, but reflecting that it is a sailor's peril to which those on the biggest and staunchest of ships are exposed almost equally with the fishermen. For was it not on the Banks and in a fog that the blow was struck which sent "La Bourgogne" to the bottom with more than four hundred souls?

STRIKES A SCHOONER AND SHEARS THROUGH HER LIKE A KNIFE

Ordinarily there is but short shrift for the helpless folks on a fishing vessel when struck by a liner. The keen prow cuts right through planking and stout oak frame, and the dissevered portions of the hull are tossed to starboard and to port, to sink before the white foam has faded from the wake of the destroying monster. They tell ghoulish tales of bodies sliced in twain as neatly as the boat itself; of men asleep in their bunks being decapitated, or waking, to find themselves struggling in the water with an arm or leg shorn off. And again, there are stories of escapes that were almost miraculous; of men thrown by the shock of collision out of the foretop of the schooner onto the deck of the steamship, and carried abroad in safety, while their partners mourned them as dead; of men, dozing in their bunks, startled suddenly by the grinding crash of steel and timbers, and left gazing wide-eyed at the gray sea lapping the side of their berths, where an instant before the tough oak skin of the schooner had been; of men stunned by some flying bit of wood, who, all unconscious, floated on the top of the hungry waves, until as by Divine direction, their inert bodies touched the side of a vagrant dory and were dragged aboard to life again. The Banks can perform their miracles of humanity as well as of cruelty.

Few forms of manual work are more exacting, involve more physical suffering and actual peril to life, than fishing with trawls. Under the happiest circumstances, with the sky clear, the sea moderately calm, and the air warm, it is arduous, muscle-trying, nerve-racking work. Pulling up half a mile of line, with hooks catching on the bottom, big fish floundering and fighting for freedom, and the dory dancing on the waves like mad, is no easy task. The line cuts the fingers, and the long, hard pull wearies the wrists until they ache, as though with inflammatory rheumatism. But when all this had to be done in a wet, chilling fog, or in a nipping winter's wind that freezes the spray in beard and hair, while the frost bites the fingers that the line lacerates, then the fisherman's lot is a bitter one.

The method of setting and hauling the trawls has been well described by Mr. John Z. Rogers, in "Outing," and some extracts from his story will be of interest to readers:

"The trawls were of cod-line, and tied to them at distances of six feet were smaller lines three feet in length, with a hook attached to the end. Each dory had six trawls, each one eighteen hundred feet long. The trawls were neatly coiled in tubs made by sawing flour barrels in two, and as fast as they were baited with pieces of herring they were carefully coiled into another tub, that they might run out quickly without snarling when being set.

"The last trawl was finished just before supper, at five o'clock. After supper the men enjoyed a half-hour smoke, then preparations were made to set the gear, as the trawls are called. The schooner got well to windward of the place where the set was to be made, and the first dory was lowered by a block and tackle. One of the men jumped into it, and his partner handed him the tubs of gear and then jumped in himself. The dory was made fast to the schooner by her painter as she drifted astern, and the other dories were put over in the same manner. When all the dories were disposed of the first one was cast off. One of the men rowed the boat before the wind while the other ran out the gear. First, he threw over a keg for a buoy, which could be seen from some distance. Fastened to the buoy-line at some sixty fathoms, or three hundred and sixty feet from the keg, was the trawl with a small anchor attached to sink it to the bottom. When this was dropped overboard the trawl was rapidly run out, and as fast as the end of one was reached it was tied to the next one, thus making a line of trawl ten thousand eight hundred feet long, with eighteen hundred hooks attached. After the schooner had sailed on a straight course a few hundred yards, the captain cast off the second dory, then along a little farther the third one, and so on till the five boats were all setting gear in parallel lines to each other. When all set this gear practically represented a fishing line over *ten miles* long with nine *thousand hooks* tied to it."

The trawls thus set were left out over night, the schooner picking up the dories and anchoring near the buoy of the first trawl. At daybreak the work of hauling in was begun:

"All the dories were made fast astern and left at the head of their respective trawls as the schooner sailed along. One of the men in each dory, after pulling up the anchor, put the trawl in the roller—a grooved wooden wheel eight inches in diameter. This was fastened to one side of the dory. The trawl was hauled in hand over hand, the heavy strain necessarily working the dory slowly along. The fish were taken off as fast as they appeared. A gaff—a stick about the size and length of a broom handle with a large, sharp hook attached—lay near at hand, and was frequently used in landing a fish over the side. Occasionally a fish would free itself from the trawl hook as it reached the surface, but the fisherman, with remarkable dexterity, would grab the gaff, and hook the victim before it could swim out of reach. What would be on the next hook was always an interesting uncertainty, for it seemed that all kinds of fish were represented. Cod and haddock were, of course, numerous, but hake and pollock struggled on many a hook. Besides these, there was the brim, a small, red fish, which is excellent fried; the cat fish, also a good pan fish; the cusk, which is best baked; the whiting, the eel, the repulsive-looking skate, the monk, of which it can almost be said that his mouth is bigger than himself, and last, but not least, that ubiquitous fish, the curse of amateur harbor fishers, the much-abused sculpin. Nor were fish alone caught on the hooks, for stones were frequently pulled up, and one dory brought in a lobster, which had been hooked by his tail. Some of the captives showed where large chunks had been bitten out of them by larger fish, and sometimes, when a hook appeared above water, there would be nothing on it but a fish head. This was certainly a case of one fish taking a mean advantage of another."

Such is the routine of trawling when weather and all the fates are propitious. But the Banks have other stories to tell—stories of men lost in the fog, drifting for long days and nights until the little keg of fresh water and the scanty store of biscuit are exhausted, and then slowly dying of starvation, alone on the trackless sea; of boats picked up in winter with frozen bodies curled together on the floor, huddled close in a vain endeavor to keep warm; of trawlers looking up from their work to see towering high above them the keen prow of an ocean grayhound, and thereafter seeing nothing that their dumb lips could tell to mortal ears. Many a story of suffering and death the men skilled in the lore of the Banks could tell, but most eloquent of all stories are those told by the figures of the men lost from the fishing ports of New England. From Gloucester alone, in 1879, two hundred and fifty fishermen were lost. In one storm in 1846 Marblehead lost twelve vessels and sixty-six men and boys. In 1894, and the first month of 1895, one hundred and twenty-two men sailing out of Gloucester, were drowned. In fifty years this little town gave to the hungry sea two thousand two hundred men, and vessels valued at nearly two million, dollars. Full of significance is the fact that every fishing-boat sets aside part of the proceeds of its catch for the widows' and orphans' fund before making the final division among the men. One of the many New England poets who have felt and voiced the pathos of life in the fishing villages, Mr. Frank H. Sweet, has told the story of the old and oft-repeated tragedy of the sea in these verses:

"THE WIVES OF THE FISHERS

"The boats of the fishers met the wind

And spread their canvas wide,

And with bows low set and taffrails wet

Skim onward side by side;

The wives of the fishers watch from shore,

And though the sky be blue,

They breathe a prayer into the air

As the boats go from view.

"The wives of the fishers wait on shore

With faces full of fright,

And the waves roll in with deafening din

Through the tempestuous night;

The boats of the fishers meet the wind

Cast up by a scornful sea;

But the fishermen come not again,

Though the wives watch ceaselessly."

CHAPTER X

THE SAILOR'S SAFEGUARDS—IMPROVEMENTS IN MARINE ARCHITECTURE—THE MAPPING OF THE SEAS—THE LIGHTHOUSE SYSTEM—BUILDING A LIGHTHOUSE—MINOT'S LEDGE AND SPECTACLE REEF—LIFE IN A LIGHTHOUSE—LIGHTSHIPS AND OTHER BEACONS—THE REVENUE MARINE SERVICE—ITS FUNCTION AS A SAFEGUARD TO SAILORS—ITS WORK IN THE NORTH PACIFIC—THE LIFE-SAVING SERVICE—ITS RECORD FOR ONE YEAR—ITS ORIGIN AND DEVELOPMENT—THE PILOTS OF NEW YORK—THEIR HARDSHIPS AND SLENDER EARNINGS—JACK ASHORE—THE SAILORS' SNUG HARBOR.

Into the long struggle between men and the ocean the last half century has witnessed the entrance of System, Science and Cooperation on the side of man. They are three elements of strength which ordinarily assure victory to the combatant who enlists them, but complete victory over the ocean is a thing never to be fully won. Build his ships as he may, man them as he will, map out the ocean highways never so precisely, and mark as he may with flaring beacons each danger point, yet in some moment of wrath the winds and the waves will rise unconquerable and sweep all the barriers, and all the edifices erected by man out of their path. To-day all civilized governments join in devices and expedients for the protection and safeguard of the mariner. Steel vessels are made unsinkable with water-tight compartments, and officially marked with a Plimsoll load line beneath which they must not be submerged. Charts of every ocean are prepared under governmental supervision by trained scientists. Myriads of lights twinkle from headland to reef all round the world. Pilots are taught to find the way into the narrowest harbors, though they can scarce see beyond the ship's jibboom, and electric-lighted buoys mark the channel, while foghorns and sirens shriek their warnings through flying scud and mist. Revenue cutters ply up and down the coast specially charged to go swiftly to the rescue of vessels in distress, and life-saving stations dot the beaches, fitted with every device for cheating the breakers of their prey. The skill of marine architects, and all the resources of Government are taxed to the utmost to defeat the wrath of Ocean, yet withal his toll of life and property is a heavy one.

Now and again men discuss the nature of courage, and try to fix upon the bravest deed of history. Doubtless *the* bravest deed has no place in history, for it must have been the act of some unknown man committed with none to observe and recount the deed. Gallantry under the stimulus of onlookers ready to cheer on the adventurer and to make history out of his exploit, is not the supremest type. Surely first among the brave, though unknown men, we must rank that navigator, who, ignorant of the compass and even of the art of steering by the stars, pressed his shallop out beyond sight of land, into the trackless sea after the fall of night. Such a one braved, beside the ordinary dangers of the deep, the uncouth and mythical terrors with which world-wide ignorance and superstition had invested it. The sea was thought to be the domain of fierce and ravenous monsters, and of gods quite as dangerous to men. Prodigious whirlpools, rapids, and cataracts, quite without any physical reason for existence, were thought to roar and roll just beyond the horizon. It is only within a few decades that the geographies have abandoned the pleasing fiction of the maelstrom, and a few centuries ago the sudden downpour of the waters at the "end of the world" was a thoroughly accepted tenet of physical geography. Yet men, adventurous and inquisitive, kept ever pushing forward into the unknown, until now there remain no strange seas and few uncharted and unlighted. The mariner of these days has literally plain sailing in comparison with his forbears of one hundred and fifty years ago.

Easily first among the sailor's safeguards is the lighthouse system. That of the United States is under the direct control of the Light House Board, which in turn is subject to the authority of the Secretary of the Treasury. It is the practice of every nation to light its own coast; though foreign vessels enjoy equal advantages thereby with the ships of the home country. But the United States goes farther. Not only does it furnish the beacons to guide foreign ships to its ports; but, unlike Great Britain and some other nations, it levies no charge upon the beneficiaries. In order that American vessels might not be hampered by the light dues imposed by foreign nations, the United States years ago bought freedom from several states for a lump sum; but Great Britain still exacts dues, a penny a ton, from every vessel passing a British light and entering a British port.

The history of the lighthouses of the world is a long one, beginning with the story of the famous Pharos, at Alexandria, 400 feet high, whose light, according to Ptolemy, could be seen for 40 miles. Pharos long since disappeared, overthrown, it is thought, by an earthquake. France possesses to-day the oldest and the most impressive lighthouse—the Corduan tower, at the mouth of the Gironde, begun in the fifteenth century. In the United States, the lighthouse system dates only from 1715, when the first edifice of this character was begun at the entrance to Boston harbor. It was only an iron basket perched on a beacon, in which were burned "fier bales of pitch and ocum," as the colonial records express it Sometimes tallow candles illuminated this pioneer light of the establishment of which announcement was made in the Boston *News*, of September 17, 1716, in this wise: "Boston. By Vertue of an Act of Assembly made in the First Year of His Majesty's Reign, For Building & Maintaining a Light House upon the Great Brewster (called Beacon Island) at the Entrance of the Harbor of Boston, in order to prevent the loss of the Lives & Estates of His Majesty's Subjects; the said Light House has been built; And on Fryday last the 14th Currant the Light was kindled; which will be very useful for all Vessels going out and coming in to the Harbor of Boston for which all Masters shall pay to the Receiver of Impost, One Peny per Ton Inwards, and another Peny Outwards, except Coasters, who are to pay Two Shillings each at their clearance Out. And all Fishing Vessels, Wood Sloops, &c. Five Shillings each by the Year."

When the United States Government was formed, with the adoption of the Constitution in 1789, there were just eight lights on the coast, namely, Portsmouth Light, N.H.; the Boston Light, mentioned above; Guerney Light, near Plymouth, Mass.; Brand Point Light, on Nantucket; Beaver Tail Light, R.I.; Sandy Hook Light; Cape Henlopen Light, Del.; and Charleston Main Light, on Morris Island, S.C. The Pacific coast, of course, was dark. So, too, was the Gulf of Mexico, though already a considerable shipping was finding its way thither. Of the multitudes of lights that gleam and sparkle in Long Island Sound or on the banks of the navigable rivers that open pathways into the interior, not one was then established. But as soon as a national government took the duty in hand, the task of lighting the mariner's pathway was pressed with vigor. By 1820 the eight lights had increased to fifty-five. To-day there are 1306 lighthouses and lighted beacons, and forty-five lightships. As for buoys, foghorns, day beacons, etc., they are almost uncounted. The board which directs this service was organized in 1852. It consists of two officers of high rank in the navy, two engineer officers of the army, and two civilians of high scientific attainments. One officer of the army and one of the navy are detailed as secretaries. The Secretary of the Treasury is *ex officio* president of the board. Each of the sixteen districts into which the country is divided is inspected by an army and a navy officer, and a small navy of lighthouse tenders perform the duty of carrying supplies and relief to the lighthouses up and down our three coasts.

MINOT'S LEDGE LIGHT

The planning of a lighthouse to stand on a submerged reef, in a stormy sea, is an engineering problem which requires extraordinary qualities of technical skill and scientific daring for its solution, while to raise the edifice, to seize the infrequent moments of low calm water for thrusting in the steel anchors and laying the heavy granite substructure on which shall rise the slender stone column that shall defy the assaults of wind and wave, demands coolness, determination, and reckless courage. Many lights have been built at such points on our coast, but the ponderous tower of Minot's Ledge, at the entrance to Boston Harbor, may well be taken as a type.

Minot's Ledge is three miles off the mouth of Boston Bay, a jagged reef of granite, wholly submerged at high tide, and showing a scant hundred yards of rock above the water at the tide's lowest stage. It lies directly in the pathway of ships bound into Boston, and over it, on even calm days, the breakers crash in an incessant chorus. Two lighthouses have reared their heads here to warn away the mariner. The first was completed in 1848, an octagonal tower, set on wrought-iron piles extending five feet into the rock. The skeleton structure was expected to offer little surface to the shock of the waves, and the wrought iron of which it was built surely seemed tough enough to resist any combined force of wind and water; but in an April gale in 1851 all was washed away, and two brave keepers, who kept the lamp burning until the tower fell, went with it. Late at night, the watchers on the shore at Cohasset, three miles away, heard the tolling of the lighthouse bell, and through the flying scud caught occasional glimpses of the light; but morning showed nothing left of the structure except twisted stumps of iron piles, bent and gnarled, as though the waves which tore them to pieces had been harder than they.

Then, for a time, a lightship tossed and tugged at its cables to warn shipping away from Minot's Ledge. Old Bostonians may still remember the gallant Newfoundland dog that lived on the ship, and, when excursion boats passed, would plunge into the sea and swim about, barking, until the excursionists would throw him tightly rolled newspapers, which he would gather in his jaws, and deliver to the lightship keepers to be dried for the day's reading. But, while the lightship served for a temporary beacon, a new tower was needed that might send the warning pencil of light far out to sea. Minot's was too treacherous a reef and too near a populous ocean highway to be left without the best guardian that science could devise. Accordingly, the present stone tower was planned and its construction begun in 1855. The problem before the designer was no easy one. The famous Eddystone and Skerryvore lighthouses, whose triumphs over the sea are related in English verse and story, were easier far to build, for there the foundation rock is above water at every low tide, while at Minot's Ledge the bedrock on which the base of the tower rests is below the level of low tide most of the year. The working season could only be from April 1 to September 15. Nominally, that is almost six months; but in the first season the sea permitted exactly 130 hours' work; in the second season 157, and in the third season, 130 hours and 21 minutes. The rest of the time the roaring surf held Minot's Ledge for its own. Nor was this all. After two years' work, the piles and débris of the old lighthouse had been cleared away, and a new iron framework, intended to be anchored in solid masonry, had been set, when up came a savage gale from the northeast; and when it cleared all was swept away. Then the spirit of the builder wavered, and he began to doubt that any structure built by men could withstand the powers of nature at Minot's Ledge. But, in time, the truth appeared. A bark, the *New Eagle*, heavy laden with cotton, had been swept right over the reef, and grounded at Cohasset. Examination showed that she had carried away in her hull the framework of the new tower. Three years' heart-trying work were necessary before the first cut stone could be laid upon the rock. In the meantime, on a great table at Cohasset, a precise model of the new tower was built, each stone cut to the exact shape, on a scale of one inch to the foot, and laid in mortar. This model completed, the soil on the hillside near by was scraped away. The granite rock thus laid bare was smoothed and leveled off into a great flat circle, and there, stone by stone, the tower was built exactly as in time it should rise in the midst of the seething cauldron of foam three miles out at sea. While the masons ashore worked at the tower, the men at the reef watched their chance, and the moment a square yard of ledge was out of water at the fall of the tide, they would leap from their boats, and begin cutting it. A circle thirty feet in diameter had to be leveled, and iron rods sunk into it as anchorages for the masonry. To do that took just three years of time, though actually less than twenty-five days of working time. From the time the first cut stone was laid until the completion of the tower, was three years and three months, though in all there were but 1102 working hours.

One keeper and three assistants guard the light over Minot's Ledge. Three miles away across the sea, now blue and smiling, now black and wrathful, they can see the little group of dwellings on the Cohasset shore which the Government provides for them, and which shelter their families. The term of duty on the rocks is two weeks; at the end of each fortnight two happy men go ashore and two grumpy ones come off; that is, if the weather permits a landing, for keepers have been stormbound for as long as seven weeks. The routine of duty is much the same in all of the lighthouses. By night there must be unceasing watch kept of the great revolving light; and, if there be other lights within reach of the keeper's glass, a watch must be kept on them as well, and any eclipse, however brief, must be noted in the lighthouse log. By day the lens must be rubbed laboriously with a dry cloth until it shines like the facets of a diamond. Not at all like the lens we are familiar with in telescopes and cameras is this scientifically contrived device. It is built up of planes and prisms of the finest flint glass, cut and assembled according to abtruse mathematical calculations so as to gather the rays of light from the great sperm-oil lamp into parallel rays, a solid beam, which, in the case of Minot's Ledge light, pierces the night to a distance of fifteen miles. On foggy days, too, the keepers must toll the fog-bell, or, if the light be on the mainland, operate the steam siren which sends its hoarse bellow booming through the gray mist to the alert ears of the sailor miles away.

The regulations do not prescribe that the keeper of a light shall hold himself ready to go to the assistance of castaways or of wrecked vessels; but, as a matter of fact, not a few of the most heroic rescues in the history of the coast have been performed by light-keepers. In the number of lives saved a woman—Ida Lewis, the keeper of the Limerock Light in Newport Harbor—leads all the rest. But there is hardly any light so placed that a boat can be launched that has not a story to tell of brave men putting out in frail boats in the teeth of a roaring gale to bring in some exhausted castaways, to carry a line to some stranded ship, or to guide some imperiled pleasure-seekers to safety.

While the building of the Minot's Ledge light had in it more of the picturesque element than attaches to the record of construction of the other beacons along the coast of the United States, there are but few erected on exposed points about which the builders could not tell some curious stories of difficult problems surmounted, or dire perils met and conquered. The Great Lakes, on which there are more than 600 light stations, offer problems of their own to the engineer. Because of the shallowness of their waters, a gale speedily kicks up a sea which old Ocean itself can hardly outdo, and they have an added danger in that during the winter they are frozen to such a depth that navigation is entirely abandoned. The lights, too, are abandoned during this season, the Lighthouse Board fixing a period in the early winter for extinguishing them and another in spring for reilluminating them. But between these dates the structures stand exposed to the tremendous pressure of such shifting floes of ice as are not found on the ocean outside of the Arctic regions. The lake lighthouse, the builders of which had most to apprehend from this sort of attack, is that at Spectacle Reef, in Lake Huron, near the Straits of Mackinaw. It is ten miles from land, standing on a limestone reef, and in the part of the lakes where the ice persists longest and moves out with the most resistless crush. To protect this lighthouse, it was necessary to build a rampart all about it, against which the ice floes in the spring, as the current moves them down into Lake Huron, are piled up in tumultuous disorder. In order to get a foundation for the lighthouse, a huge coffer-dam was built, which was launched like a ship, towed out to the reef and there grounded. When it was pumped out the men worked inside with the water surrounding them twelve to fourteen feet above their heads.

Twenty months of work, or three years in time, were occupied in erecting this light. Once in the spring, when the keepers returned after the closed season to prepare for the summer's navigation, they found the ice piled thirty feet against the tower, and seventy feet above the doorway, so that they were compelled, in order to enter the lighthouse, to cut through a huge iceberg of which it was the core.

The Spectacle Reef light, like that at Minot's Ledge, is a simple tower of massive masonry, and this is the approved design for lighthouses exposed to very heavy strain from waves or ice. A simpler structure, used in tranquil bays and in the less turbulent waters of the Gulf, is the "screw-pile" lighthouse, built upon a skeleton framework of iron piling, the piles having been so designed that they bore into the bed of the ocean like augers on being turned. The "bug-light" in Boston Harbor, and the light at the entrance to Hampton Roads are familiar instances of this sort of construction. For all their apparent lightness of construction, they are stout and seaworthy, and in their erection the builders have often had to overcome obstacles and perils offered by the sea scarcely less savage than those overcome at Minot's Ledge. Indeed, a lighthouse standing in its strength, perhaps rising out of a placid summer sea, or towering from a crest of rock which it seems incredible the sea should have ever swept, gives little hint to the casual observer of the struggle that brave and skilful men had to go through with before it could be erected. The light at Tillamook Rock, near the mouth of the Columbia River, offers a striking illustration of this. It is no slender shaft rising from a tumultuous sea, but a spacious dwelling from which springs a square tower supporting the light, the whole perched on the crest of a small rock rising precipitously from the sea to the height of some forty feet. Yet, sturdy and secure as the lighthouse now looks, its erection was one of the hardest tasks that the board ever undertook. So steep are the sides of Tillamook Rock that to land upon it, even in calm weather, is perilous, and the foreman of the first party that went to prepare the ground for the light was drowned in the attempt. Only after repeated efforts were nine men successfully landed with tools and provisions. Though only one mile from shore they made provision for a prolonged stay, built a heavy timber hut, bolting it to the rock, and began blasting away the crest of the island to prepare foundations for the new lighthouse. High as they were above the water, the sea swept over the rock in a torrent when the storms raged. In one tempest the hut was swept away and the men were barely able to cling to the rock until the waves moderated. That same night an English bark went to pieces under the rock, so near that the workmen above, clinging for dear life to their precarious perch, could hear the shouts of her officers giving their commands. A bonfire was kindled, in hope of warning the doomed sailors of their peril, but it was too late, for the ship could not be extricated from her position, and became a total wreck, with the loss of the lives of twenty of her company. To-day a clear beam of light shines out to sea, eighteen miles from the top of Tillamook, and only the criminally careless captain can come near enough to be in any danger whatsoever. Such is one bit of progress made in safeguarding the sea.

More wearing even than life in a lighthouse is that aboard the lightships, of which the United States Government now has forty-five in commission. The lightship is regarded by the Government as merely a makeshift, though some of them have been in use for more than a quarter of a century. They are used to mark shoals and reefs where it has thus far been impossible to construct a lighthouse, or obstructions to navigation which may be but temporary. While costing less than lighthouses, they are not in favor with the Lighthouse Board, because the very conditions which make a light most necessary, are likely to cause these vessels to break from their moorings and drift away, leaving their post unguarded. Their keepers suffer all the discomforts of a sailor's life and most of its dangers, while enjoying none of its novelty and freedom. The ships are usually anchored in shoal water, where the sea is sure to run high, and the tossing and rolling of the craft makes life upon it insupportable. They are always farther out to sea than the lighthouses, and the opportunities for the keepers to get ashore to their families are correspondingly fewer. In heavy storms their decks are awash, and their cabins dripping; the lights, which must be watched, instead of being at the top of a firm, dry tower, are perched on reeling masts over which the spray flies thick with every wave, and on which is no shelter for the watcher. During long weeks in the stormy season there is no possible way of escaping from the ship, or of bringing supplies or letters aboard, and the keepers are as thoroughly shut off from their kind as though on a desert island, although all day they may see the great ocean liners steaming by, and through their glasses may be able to pick out the roofs of their cottages against the green fields far across the waves.

WHISTLING BUOY

Less picturesque than lighthouses and lightships, and with far less of human interest about them, are the buoys of various sorts of which the Lighthouse Board has more than one thousand in place, and under constant supervision.

Yet, among the sailor's safeguards, they rank near the head. They point out for him the tortuous pathway into different harbors; with clanging bell or dismal whistle, they warn him away from menacing shallows and sunken wrecks. The resources of science and inventive genius have been drawn upon to devise ways for making them more effective. At night they shine with electric lights fed from a submarine cable, or with steady gas drawn from a reservoir that needs refilling only three or four times a year. If sound is to be trusted rather than light, recourse is had to a bell-buoy which tolls mournfully as the waves toss it about above the danger spot, or to a whistling buoy which toots unceasingly a locomotive whistle, with air compressed by the action of the waves. The whistling buoy is the giant of his family, for the necessity for providing a heavy charge of compressed air compels the attachment to the buoy of a tube thirty-two feet or more deep, which reaches straight down into the water. The sea rising and falling in this, as the buoy tosses on the waves, acts as a sort of piston, driving out the air through the whistle, as the water rises, admitting more air as it falls.

Serving a purpose akin to the lighthouses, are the post and range-lights on the great rivers of the West. Very humble devices, these, in many instances, but of prodigious importance to traffic on the interior waterways. A lens lantern, hanging from the arm of a post eight or ten feet high, and kept lighted by some neighboring farmer at a cost of $160 a year, lacks the romantic quality of a lighthouse towering above a hungry sea, but it is because there are nearly two thousand such lights on our shallow and crooked rivers that we have an interior shipping doing a carrying trade of millions a year, and giving employment to thousands of men.

Chief among the sailors' safeguards is the service performed by the United States revenue cutters. The revenue cutter service, like the lighthouse system, was established very shortly after the United States became a nation by the adoption of the Constitution. Its primary purpose, of course, is to aid in the enforcement of the revenue laws and to suppress smuggling. The service, therefore, is a branch of the Treasury Department, and is directly under the charge of the Secretary of the Treasury. In the course of years, however, the revenue cutter service has extended its functions. In time of war, the cutters have acted as adjuncts to the navy, and some of the very best armed service on the high seas has been performed by them. Piracy in the Gulf of Mexico was largely suppressed by officers of revenue cutters, and pitched battles have more than once been fought between small revenue cutters and the pirates of the Louisiana and the Central American coasts.

But the feature of the service which is of particular pertinence to our story of American ships and sailors, is the part that it has taken in aiding vessels that were wrecked, or in danger of being wrecked. Many years ago, the Secretary of the Treasury directed the officers of the revenue marine to give all possible aid to vessels in distress wherever encountered. Perhaps the order was hardly necessary. It is the chiefest glory of the sailor, whether in the official service, or in the merchant marine, that he has never permitted a stranger ship to go unaided to destruction, if by any heroic endeavor he could save either the ship or her crew. The annals of the sea are full of stories of captains who risked their own vessels, their own lives, and the lives of their people, in order to take castaways from wrecked or foundering vessels in a high sea. But the records of the revenue marine service are peculiarly fruitful of such incidents, because it was determined some thirty years ago that cutters should be kept cruising constantly throughout the turbulent winter seasons for the one sole purpose of rendering aid to vessels in distress. In these late years, when harbors are thoroughly policed, and when steam navigation has come to dominate the ocean, there is little use for the revenue cutter in its primary quality of a foe to smugglers. People who smuggle come over in the cabins of the finest ocean liners, and the old-time contraband importer, of the sort we read of in "Cast Up By The Sea," who brings a little lugger into some obscure port under cover of a black night, has entirely disappeared.

A duty which at times has come very near to true war service, has been the enforcement of the *modus vivendi* agreed upon by Great Britain and the United States, as a temporary solution of the problem of the threatened extinction of the fur-bearing seals. This story of the seal "fishery," and the cruel and wholesale slaughter which for years attended it, is one of the most revolting chapters in the long history of civilized man's warfare on dumb animals. It is to be noted that it is only the civilized man who pursues animals to the point of extinction. The word "savage" has come to mean murderous, bloodthirsty, but the savages of North America hunted up and down the forests and plains for uncounted centuries, living wholly on animal food, finding at once their livelihood and their sport in the chase, dressing in furs and skins, and decking themselves with feathers, but never making such inroads upon wild animal life as to affect the herds and flocks. Civilized man came with his rifles and shot-guns, his eagerness to kill for the sake of killing, his cupidity, which led him to ignore breeding-seasons, and seek the immediate profit which might accrue from a big kill, even though thereby that particular form of animal life should be rendered extinct. In less than forty years after his coming to the great western plains, the huge herds of buffalo had disappeared. The prairie chicken and the grouse became scarce, and fled to the more remote regions. Of lesser animal life, the woods and fields in our well-settled states are practically stripped bare. A few years ago, it became apparent that for the seals of the North Pacific ocean and Bering Sea, early extinction was in store. These gentle and beautiful animals are easily taken by hunters who land on the ice floes, where they bask by the thousands, and slaughter them right and left with heavy clubs. The eager demand of fashionable women the world over for garments made of their soft, warm fur, stimulated pot-hunters to prodigious efforts of murder. No attention was given to the breeding season, mothers with young cubs were slain as ruthlessly as any. Schooners and small steamers manned by as savage and lawless men as have sailed the seas since the days of the slave-trade, put out from scores of ports, each captain eager only to make the biggest catch of the year, and heedless whether after him there should be any more seals left for the future. This sort of hunting soon began to tell on the numbers of the hapless animals, and the United States Government sent out a party of scientific men in the revenue cutter "Lincoln," to investigate conditions, particularly in the Pribylof Islands, which had long been the favorite sealing ground. As a result of this investigation, the United States and Great Britain entered into a treaty prohibiting the taking of seals within sixty miles of these islands, thus establishing for the animals a safe breeding-place. The enforcement of the provisions of this treaty has fallen upon the vessels of the revenue service, which are kept constantly patrolling the waters about the islands, boarding vessels, counting the skins, and investigating the vessel's movements. It has been a duty requiring much tact and firmness, for many of the sealers are British, and the gravest international dissension might have arisen from any unwarrantable or arbitrary interference with their acts. The extent of the duty devolving upon the cutters is indicated by some figures of their work in a single year. The territory they patrolled covered sixty degrees of longitude and twenty-five of latitude, and the cruising distance of the fleet was 77,461 miles. Ninety-four vessels were boarded and examined, over 31,000 skins counted, and four vessels were seized for violation of the treaty. In the course of this work, the cutters engaged in it have performed many useful and picturesque services. On one occasion it fell to one of them to go to the rescue of a fleet of American whalers who, nipped by an unusually early winter in the polar regions, were caught in a great ice floe, and in grave danger of starving to death. The men from the cutters hauled food across the broad expanse of ice, and aided the imprisoned sailors to win their freedom. The revenue officers, furthermore, have been to the people of Alaska the respected representatives of law and order, and in many cases the arbiters and enforcers of justice. Along the coast of Alaska live tribes of simple and ignorant Indians, who were for years the prey of conscienceless whites, many of whom turned from the business of sealing, when the two Governments undertook its regulation, to take up the easier trade of fleecing the Indians. The natives were all practised trappers and hunters, and as the limitations upon sealing did not apply to them, they had pelts to sell that were well worth the buying. Ignorant of the values of goods, eager for guns and glittering knives, and always easily stupefied with whisky, the Indians were easy prey to the sea traders. For a gun of doubtful utility, or a jug of fiery whisky, the Indian would not infrequently barter away the proceeds of a whole year of hunting and fishing, and be left to face the winter with his family penniless. It has been the duty of the officers of the revenue cutters serving on the North Pacific station to suppress this illicit trade, and to protect the Indians, as far as possible, from fraud and extortion. The task has been no easy one, but it has been discharged so far as human capacity would permit, so that the Alaska Indians have come to look upon the men wearing the revenue uniform as friends and counselors, while to a great extent the semi-piratical sailors who infested the coast have been driven into other lines of dishonest endeavor. Perhaps not since the days of Lafitte and the pirates of Barataria has any part of the coast of the United States been cursed with so criminal and abandoned a lot of sea marauders as have for a decade frequented the waters off Alaska, the Pribylof Islands, and the sealing regions. The outlawry of a great part of the seal trade, and the consequent heavy profits of those who are able to make one or two successful cruises uncaught by officers of the law, have attracted thither the reckless and desperate characters of every sea, and with these the revenue cutters have to cope. Yet so diversified are the duties of this service that the revenue officers may turn from chasing an illicit sealer to go to the rescue of whalers nipped in the ice, or may make a cruise along the coast to deliver supplies from the Department of Education to mission schools along Bering Sea and the Arctic Ocean, or to carry succor to a party of miners known to be in distress. The rapid development of Alaska since the discoveries of gold has greatly added to the duties of this fleet.

REVENUE CUTTER

The revenue service stands midway between the merchant service and the navy. It may almost be said that the officers engaged in it suffer the disadvantages of both forms of sea service without enjoying the advantages of either. Unlike navy officers, they do not have a "retired list" to look forward to, against the time when they shall be old, decrepit, and unfit for duty. Congress has, indeed, made provision for placing certain specified officers on a roll called "permanent waiting orders," but this has been but a temporary makeshift, and no officer can feel assured that this provision will be made for him. Promotion, too, while quite as slow as in the navy, is limited. The highest officer in the service is a captain, his pay $2500 a year— but a sorry reward for a lifetime of arduous labor at sea, during which the officer may have been in frequent peril of his life, knowing all the time that for death in the discharge of duty, the Government will pay no pension to his heirs unless the disaster occurred while he was "cooperating with the navy." In one single year the records of the revenue service show more than one hundred lives saved by its activity, without taking into consideration those on vessels warned away from dangerous points by cutters. Yet neither in pay, in provision for their old age, or for their families in case of death met in the discharge of duty, are the revenue officers rewarded by the Government as are navy officers, while public knowledge and admiration for the service is vastly less than for the navy. It is a curious phenomenon, and yet one as old at least as the records of man, that the professional killer— that is to say, the officer of the army or navy—has always been held in higher esteem socially, and more lavishly rewarded, than the man whose calling it is to save life.

To a very considerable degree the life-saving service of the United States is an outgrowth of the revenue marine. To sojourners by the waterside, on the shores of either ocean or lake, the trim little life-saving stations are a familiar sight, and summer pleasure-seekers are entertained with the exhibition drills of the crews in the surf. It is the holiday side of this service as a rule that the people chiefly know, but its records show how far from being all holiday pleasure it is. In 1901 the men of the life-saving corps were called to give aid to 377 wrecked ships. Of property in jeopardy valued at $7,354,000, they saved $6,405,035 worth. Of 93,792 human beings in peril of death in the waters, all save 979 were saved. These are the figures relating only to considerable shipwrecks, but as life-saving stations are established at nearly every harbor's mouth, and are plentiful about the pleasure cruising grounds of yachts and small sailboats, hundreds of lives are annually saved by the crews in ways that attract little attention. In 1901 the records show 117 such rescues.

The idea of the life-saving service originated with a distinguished citizen of New Jersey, a State whose sandy coast has been the scene of hundreds of fatal shipwrecks. In the summer of 1839 William A. Newell, a young citizen of that State, destined later to be its Governor, stood on the beach near Barnegat in a raging tempest, and watched the Austrian brig "Count Perasto" drift onto the shoals. Three hundred yards from shore she struck, and lay helpless with the breakers foaming over her. The crew clung to the rigging for a time, but at last, fearing that she was about to go to pieces, flung themselves into the raging sea, and strove to swim ashore. All were drowned, and when the storm went down, the dead bodies of thirteen sailors lay strewn along the beach, while the ship itself was stranded high and dry, but practically unhurt, far above the water line.

"The bow of the brig being elevated and close to the shore after the storm had ceased," wrote Mr. Newell, in describing the event long years after, "the idea was forced quickly upon my mind that those unfortunate sailors might have been saved if a line could have been thrown to them across the fatal chasm. It was only a short distance to the bar, and they could have been hauled ashore in their small boat through, or in, the surf.... I instituted experiments by throwing light lines with bows and arrows, by rockets, and by a shortened blunderbuss with ball and line. My idea culminated in complete success, however, by the use of a mortar, or a carronade, and a ball and line. Then I found, to my great delight, that it was an easy matter to carry out my desired purpose."

Shortly after interesting himself in this matter Mr. Newell was elected to Congress, and there worked untiringly to persuade the national Government to lend its aid to the life-saving system of which he had conceived the fundamental idea. In 1848 he secured the first appropriation for a service to cover only the coast of New Jersey. Since then it has been continually extended until in 1901 the life-saving establishment embraced 270 stations on the Atlantic, Pacific, and lake coasts. The appropriation for the year was $1,640,000. For many years the service was a branch of the revenue marine, and when in 1878 it was made a separate bureau, the former chief of the revenue marine bureau was put at its head. The drill-masters for the crews are chosen from the revenue service, as also are the inspectors.

LAUNCHING A LIFEBOAT THROUGH THE SURF

The methods of work in the life-saving service have long been familiar, partly because at each of the recurring expositions of late years, the service has been represented by a model station and a crew which went daily through all the operations of shooting a line over a stranded ship, bringing a sailor ashore in the breeches-buoy or the life-car, and drilling in the non-sinkable, self-righting surf-boat. Along the Atlantic coast the stations are so thickly distributed that practically the whole coast from Sandy Hook to Hatteras is continually under patrol by watchful sentries. Night and day, if the weather be stormy or threatening, patrolmen set out from each station, walking down the beach and keeping a sharp eye out for any vessel in the offing. Midway between the stations they meet, then each returns to his own post. In the bitter nights of winter, with an icy northeaster blowing and the flying spray, half-frozen, from the surf, driven by the gale until it cuts like a knife, the patrolman's task is no easy one. Indeed, there is perhaps no form of human endeavor about which there is more constant discomfort and positive danger than the life-saving service. It is the duty of the men to defy danger, to risk their lives whenever occasion demands, and the long records of the service show uncounted cases of magnificent heroism, and none of failure in the face of duty.

A form of seafaring which still retains many of the characteristics of the time when Yankee sailors braved all seas and all weather in trig little wooden schooners, is the pilot service at American ports, and notably at New York. Even here, however, the inroads of steam are beginning to rob the life of its old-time picturesqueness, though as they tend to make it more certain that the pilot shall survive the perils of his calling, they are naturally welcomed. Under the law every foreign vessel entering an American port must take a pilot. If the captain thinks himself able to thread the channel himself, he may do so; but nevertheless he has to pay the regular pilot fee, and if the vessel is lost, he alone is responsible, and his owners will have trouble with the insurance companies. So the law is acquiesced in, perhaps not very cheerfully, and there have grown up at each American port men who from boyhood have studied the channels until they can thread them with the biggest steamship in the densest fog and never touch bottom. New York as the chief port has the largest body of pilots, and in the old days, before the triumph of steam, had a fleet of some thirty boats, trim little schooners of about eighty tons, rigged like yachts, and often outsailing the best of them. In those days the rivalry between the pilots for ships was keen and the pilot-boats would not infrequently cruise as far east as Sable Island to lay in wait for their game. That was in the era of sailing ships and infrequent steamers, and it was the period of the greatest mortality among the pilots; for staunch as their little boats were, and consummate as was their seamanship, they were not fitted for such long cruises. The marine underwriters in those days used to reckon on a loss of at least one pilot-boat annually. Since 1838 forty-six have been lost, thirteen going down with all on board. In late years, however, changes in the methods of pilotage have greatly decreased the risks run by the boats. When the great ocean liners began trying to make "record trips" between their European ports and Sandy Hook, their captains became unwilling to slow up five hundred miles from New York to take a pilot. They want to drive their vessels for every bit of speed that is in them, at least until reported from Fire Island. The slower boats, the ocean tramps, too, look with disfavor on shipping a pilot far out at sea, for it meant only an idler aboard, to be fed until the mouth of the harbor was reached. So the rivalry between the pilots gave way to cooperation. A steamer was built to serve as a station-boat, which keeps its position just outside New York harbor, and supplies pilots for the eight boats of the fleet that cruise over fixed beats a few score miles away. But this change in the system has not so greatly reduced the individual pilot's chance of giving up his life in tribute to Neptune, for the great peril of his calling—that involved in getting from his pilot-boat to the deck of the steamer he is to take in—remains unabated.

THE EXCITING MOMENT IN THE PILOT'S TRADE

Professional pride no less than hope of profit makes the pilot take every imaginable risk to get to his ship. He draws no regular salary, but his fee is graduated by the draft of the vessel he pilots. When a ship is sighted coming into port, the pilot-boat makes for her. If she has a blue flag in her rigging, half way up, by day, she has a pilot aboard. At night, the pilot-boats show a blue flare, by way of query. If the ship makes no answer, she is known to be supplied, and passes without slowing up; but if in response to signal she indicates that she is in need of a pilot, the exciting moment in the pilot's trade is at hand. Perhaps the night is pitchy dark, with a gale blowing and a heavy sea on: but the pilot slips on his shore clothes and his derby hat—it is considered unprofessional to wear anything more nautical—and makes ready to board. The little schooner runs up to leeward of where the great liner, with her long rows of gleaming portholes, lies rolling heavily in the sea. Sharp up into the wind comes the midget, and almost before she has lost steerage way a yawl is slid over the side, the pilot and two oarsmen tumble into it, and make for the side of the steamship. To climb a rope-ladder up the perpendicular face of a precipice thirty feet high on an icy night is no easy task at best; but if your start is from a boat that is being tossed up and down on a rolling sea, if your precipice has a way of varying from a strict perpendicular to an overhanging cliff, and then in an instant thrusting out its base so that the climber's knees and knuckles come with a sharp bang against it, while the next moment he is dropped to his shoulders in icy sea-water, the difficulties of the task are naturally increased. The instant the pilot puts his feet on the ladder he must run up it for dear life if he would escape a ducking, and lucky he is if the upward roll does not hurl him against the side of the ship with force enough to break his hold and drop him overboard. Sometimes in the dead of winter the ship is iced from the water-line to the rail, and the task of boarding is about equivalent to climbing a rolling iceberg. But whatever the difficulty, the pilot meets and conquers it—or else dies trying. It is all in the day's work for them. Accidents come in the form of boats run down by careless steamers, pilots crushed against the side or thrown into the sea by the roll of the vessel, the foundering of the pilot-boat or its loss on a lee shore; but still the ranks of the pilots are kept full by the admission to a long apprenticeship of boys who are ready to enter this adventurous and arduous calling. Few occupations require a more assiduous preparation, and the members of but few callings are able to guard themselves so well against the danger of over-competition. Nevertheless the earnings of the pilots are not great. They come under the operation of the rule already noted, that the more dangerous a calling is, the less are its rewards. Three thousand dollars a year is a high income for a pilot sailing out of New York harbor, and even this is decreasing as the ships grow bigger and fewer. Nor can he be at all certain as to what his income will be at any time, for the element of luck enters into it almost as much as into gambling. For weeks he may catch only small ships, or, the worst ill-luck that can befall a pilot, he may get caught on an outbound ship and be carried away for a six weeks' voyage, during which time he can earn nothing. But the pilot, like the typical sailor of whatever grade, is inured to hard luck and accustomed to danger.

Such are some of the safeguards which modern science and organization have provided for the sailor in pursuit of his always hazardous calling. Many others of course could be enumerated. The service of the weather bureau, by which warning of impending storms is given to mariners, is already of the highest utility. The new invention of wireless telegraphy, by which a ship at sea may call for aid from ashore, perhaps a thousand miles away, has great possibilities. Modern marine architecture is making steamships almost unsinkable, more quickly responsive to their helms, more seaworthy in every way. Perhaps with the perfection of the submarine boat, ships, instead of being tossed on the boisterous surface of the waves, may go straight to their destination through the placid depths of ocean. But whatever the future may bring, the history of the American sailor will always bear evidence that he did not wait for the perfection of safety devices to wrest from the ocean all that there was of value in the conquest; that no peril daunted him, nor was any sea, however distant, a stranger to his adventurous sail.

Much has been said and written of the improvidence of the sailor, of his profligacy when in port, his childlike helplessness when in the hands of the landsharks who haunt the waterside streets, his blind reliance upon luck to get him out of difficulties, and his utter indifference to all precautionary provisions for the proverbial rainy day. Perhaps the sailor has been getting a shade the worse of it in the literature on this subject, for he, himself, is hardly literary in his habits, and has not been able to tell his own story. The world has heard much of the jolly Jack Tars who spend in a few days' revel in waterside dives the whole proceeds of a year's cruise; but it has heard less of the shrewd schemes which are devised for fleecing poor Jack, and applied by every one with whom he comes in contact, from the prosperous owner who pays him off in orders that can only be conveniently cashed at some outfitter's, who charges usurious rates for the accommodation, down to the tawdry drab who collects advance money on account of half a dozen sailor husbands. The seaman landing with money in his pocket in any large town is like the hapless fish in some of our much-angled streams. It is not enough to avoid the tempting bait displayed on every side. So thick are the hooks and snares that merely to swim along, intent on his own business, is likely to result sooner or later in his being impaled on some cruel barb. Not enough has been said, either, of the hundreds of American lads who shipped before the mast, made their voyages around Cape Horn and through all the Seven Seas, resisted the temptations of the sailors' quarters in a score of ports, kept themselves clean morally and physically, and came, in time, to the command and even the ownership of vessels. Among sailors, as in other callings, there are the idle and the industrious apprentices, and the lesson taught by Hogarth's famous pictures is as applicable to them that go down to the sea in ships as to the workers at the loom. It is doubtful, too, whether the sailor is either more gullible or more dissolute when in port than the cowboy when in town for a day's frolic, or the miner just in camp with a pocket full of dust, after months of solitude on his claim. Men are much of a sort, whatever their calling. After weeks of monotonous and wearing toil, they are apt to go to extremes when the time for relaxation comes. Men whose physical natures only are fully developed seek physical pleasures, and the sailor's life is not one to cultivate a taste for the quieter forms of recreation.

But the romance that has always surrounded the sailor's character, his real improvidence, and his supposedly unique simplicity have, in some slight degree, redounded to his advantage. They have led people in all lands to form organizations for his aid, protection, and guidance, hospitals to care for him in illness, asylums and homes to provide for the days of his old age and decrepitude. Best known of all these charitable institutions for the good of Jack Tar is the Sailor's Snug Harbor, whose dignified buildings on Staten Island look out across the finest harbor in the world to where New York's tall buildings tower high above the maintop-gallant mast of the biggest ship ever built. This institution, founded just one hundred years ago by the will of Captain Robert R. Randall, himself an American sailor of the old type, who amassed his fortune trading to all the countries on the globe, now has an income of $400,000 annually, and cares for 900 old sailors, each of whom must have sailed for at least five years under the American flag.

A new chapter in the story of the American sailor is opening as this book is closed. The period of the decadence of the American merchant marine is clearly ended, and everything gives assurance that the first quarter of this new century will do as much toward re-establishing the United States flag on the high seas as the first quarter of the nineteenth century did toward first putting it there. As these words are being written, every shipyard in the United States is busy, and some have orders that will tax their capacity for three years to come. New yards are being planned and small establishments, designed only to build pleasure craft, are reaching out after greater things. The two biggest steamships ever planned are building near New London, where four years ago was no sign of shipyard or factory. The Great Lakes and the Pacific coast ring with the sound of the steel ship-builder's hammer.

But will the American sailor share in the new life of the American ship? The question is no easy one to answer. Modern shipping methods offer little opportunity for ambitious lads to make their way from the forecastle to the owner's desk. The methods by which the Cleavelands, Crowninshields, Lows, and their fellows in the early shipping trade won their success, have no place in modern economy. As I write, the actual head of the greatest shipping concern the world has ever known, is a Wall Street banker, whose knowledge of the sea was gained from the deck of a private steam yacht or the cabin *de luxe* of a fast liner, and who has applied to the shipping business only the same methods of stock manipulating that made him the greatest railroad director in the world before he thought to control the ocean as well. With steam, the sailor has become a mere deckhand; the captain a man of business and a disciplinarian, who may not know the names of the ropes on a real ship; the owner a corporation; the voyages mere trips to and fro between designated ports made with the regularity and the monotony of a sleeping-car's trips between Chicago and San Francisco. Until these conditions shall materially change, there is little likelihood that the sea will again attract restless, energetic, and ambitious young Americans. Men of the type that we have described in earlier chapters of this book do not adopt a life calling that will forever keep them in subordinate positions, subject to the whims and domination of an employing corporation. A genial satirist, writing of the sort of men who became First Lords of the Admiralty in England, said:

> "Mind your own business and never go to sea,
>
> And you'll come to be the ruler of the Queen's navee."

Perhaps a like situation confronts the American merchant marine in its new development.